C0-AKP-951

*Called by the Wild*

# Called by the Wild

*The Autobiography of a Conservationist*

Raymond F. Dasmann

WITH A FOREWORD BY
Paul R. Ehrlich

UNIVERSITY OF CALIFORNIA PRESS

University of California Press
Berkeley and Los Angeles, California

University of California Press, Ltd.
London, England



Library of Congress Cataloging-in-Publication Data

Dasmann, Raymond Fredric, 1919–.
Called by the Wild: The Autobiography of a Conservationist/
Raymond F. Dasmann.
p. cm.
Includes bibliographical references (p. ) and index.
ISBN 0–520–22978–9 (cloth : alk. paper).
1. Dasmann—Raymond F., 1919–. 2. Conservationists—
United States—Biography. I. Title.

QH31.D37 A3 2002
333.7'2'092—dc21 2001027683
[B]

Manufactured in the United States of America
10 09 08 07 06 05 04 03
10 9 8 7 6 5 4 3 2 1

The paper used in this publication meets the minimum requirements of ANSI/NISO Z39.48–1992 (R 1997) (*Permanence of Paper*).

*To my wife, Elizabeth, whose willingness to move with me to the ends of the earth made this possible, and to our daughters, without whom this book would not have been finished.*

# CONTENTS

# FOREWORD

Paul R. Ehrlich

This graceful memoir brought me almost more nostalgia than an aging environmentalist can deal with. The biggest event in my life, even though I was thirteen years old when it ended, was the Second World War. I wanted badly to serve—the foolishness of a young boy. Subsequently, I often wondered what it would have been like to have actually seen combat, and I have read many wartime accounts. But none of them have impressed me as much as Ray Dasmann's story, perhaps because he's always been one of my heroes and shaped many of my attitudes. He was under fire in New Guinea. I never got there until 1965, when I was in the South Pacific, doing research on butterflies and visiting famous battlefields. I can remember chasing a specimen into the bush near Lae and finding a corroded tray of Nambu machine-gun ammunition. I couldn't help wondering what it would have been like to share the dense rain forest with people determined to kill me. Ray's story of cowering terrified in a foxhole while there was a "constant whine and snap of rifle or machine-gun bullets passing nearby" gave me one perspective on that. He could not see his enemies because they were hidden in the jungle—which, remembering my own experience in New Guinea jungles, I found all too understandable.

And, of course, Ray's childhood, and much of his life, was spent in central California—the land of milk and honey that is my adopted home. But Ray's stories of the San Francisco Bay area before I arrived (and of his early days in the forest service) made me wish I'd been able to flee the awful East even earlier.

Ray came into the environmental movement from game management—in some ways the same route used by Aldo Leopold. He got his feet wet in the politics of deer; there were too many for the tastes of farmers, who didn't like them eating their crops, and too few for hunters, who wanted a big supply to blow away and eat. The deer he studied had populations much larger than their range could support, but trying to introduce a doe hunt to restore some balance was anathema to the hunting community. Today the hunters would be joined by the "animal rights" movement. They would rather have natural ecosystems ruined and many populations of their nonhuman inhabitants driven to extinction than allow a hunter to kill an animal whose public image was created by the cartoon movie *Bambi.* Of course, I'm sympathetic with some of the goals of the animal rights movement—protecting animals from the "fur trade," ending some obnoxious practices in animal agriculture, and guarding against superfluous use of living animals in research (some of it is not essential). But, for example, when the movement favors extermination of elephants over sustainable harvesting by rich but not-too-bright hunters (with great benefits to local villagers), as it has in trying to sabotage the "Campfire" program in Zimbabwe, I'm on the other side.

So, I'm sure would be Ray, who had extensive experience on the magic continent of Africa. Other places where we have both learned to appreciate foreign peoples, faunas, floras, and conservation situations include Australia, Costa Rica, the Leeward Islands, and Malaysia. In 1979 he saw the destruction of the tropical forests of Malaysia for oil palm agriculture. That really struck a note for me, because Anne and I did research on butterflies in those magnificent forests, aided by aboriginal people, in 1966. In 1997 we visited the area again and were appalled at the extent to which Southeast Asia had been converted into oil palm biological deserts.

Ray is just enough older than I to have had the privilege of meeting some of the conservationists that influenced my thinking when I was an undergraduate at the University of Pennsylvania. He actually knew Bill Vogt, whose ideas (along with those of Fairfield Osborne), Ray correctly implies, formed part of the basis of my *Population Bomb.* He was in on the very start of the modern environmental movement, working with the legendary Frank Fraser Darling at the Conservation Foundation in Washington. His own 1965 book, *The Destruction of California,* published just three years after Rachel Carson's *Silent Spring,* was a pioneering monument in what was soon to be an explosion of books on the environment. Later, as he recounts, he became familiar with the ups and downs of international environmental politics. Things haven't changed much. Today advances like the Montreal Ozone Protocol contrast with foot-dragging on greenhouse agreements and the stupidity of George W. Bush's assault on U.S. aid to women in poor countries needing access to family planning.

There are many treasures in this wonderful book, and some tragedy too, but I'll leave you to discover that for yourself. Like most naturalist/environmentalists, Ray early recognized the major role played by the human population explosion in the deterioration of Earth's environment. He was there way ahead of the *Population Bomb.* And by 1971 he had coined what he called the first law of the environment, "no matter how bad you think things are—the total reality is much worse." It's sadly even more true today than it was then.

# ACKNOWLEDGMENTS

I seem to have spent years getting this book together and doubt that I could have succeeded without the help of my daughters Sandra Dasmann and Marlene Dasmann. They encouraged me to keep on writing even though I was having increasing problems and could only move slowly with handwriting and typing and could not use the computer at all. Sandra did virtually all the initial typing and editing of the manuscript and helped me write several difficult passages. Marlene stepped in toward the end to assist with chronology and linking explanatory paragraphs. To say I am grateful for this is an understatement. But I do tend to specialize in understatements.

My thanks also to Jim Clark, the director of the University of California Press, who read the first rough draft of this book, and to my editor, Mari Coates, who stayed with it through several revisions.

My appreciation to the cheerful group at Jade Mountain Health Center, especially my Breema teacher, Susan Mankowski, who helped put me back together after the death of Elizabeth, and to Cynthia Lester, who provided the psychic connection to the world beyond the dark barrier of death.

So many friends and colleagues have helped me over the years that I cannot name them all, but my gratitude for their assistance is boundless.

# Introduction

> Another damned, thick, square book! Always scribble, scribble, scribble, scribble! Eh! Mr. Gibbon?
>
> Remarks of the duke of Gloucester to Edward Gibbon on receiving a copy of his *Decline and Fall of the Roman Empire* (1781)

Any fool can see that the sun rises and sets. But it doesn't. We rotate around it. Getting that fact straight took a lot of observations and arguments. Yet we will no doubt go on forever talking about sunrises and sunsets since that is the way we see things. This book is about the way I see things, which is not necessarily what another person at the same time and place would have seen. You really can't separate the observer from the observed. To understand my ideas you should know something about my background, and then you may draw other concepts from the same sets of data. Scientists have been doing that since Aristotle or before, which is one reason why they argue so much.

This book is not just an autobiography, though it starts out that way. My concern is for the issues, problems, and challenges we are all wrestling with today, particularly those with which I have had some close involvement. I don't want to tell you all about my life. What has happened to me over the years has been a matter of opportunism, chance, or fate. I had no intention of ever becoming a professor or a writer of books. But it happened. I thought I wanted to be out in the woods, studying wildlife. I still do.

Most of my working career has been spent in some sort of professorial role at several different universities. Being a professor has certain advantages. You are forced to go through a sort of "final exam" or reality check every time you give a lecture. My students have always been quick to tell me when I was wrong, and the interaction back and forth has made it possible for me to learn from them. Many of them had extensive experience in the "real world" before they returned to academia to pick up a degree.

In a sense I have always been torn between two conflicting sides of my nature, that of a rebel and that of a hermit. A rebel seeks to be part of a rebellion aimed at changing the status quo. A hermit withdraws from society and seeks solitude. A rebel is usually an activist, but except for brief periods I have not fitted into that role. I am not inclined toward demonstrating, organizing, getting out the vote, raising money, or any of the many abilities an activist should have. And yet a professor cannot live like a hermit since he has to be out there in public encouraging students to think and learn. Still, whenever I can, I seek solitude, or the company of those nonhuman species that are our fellow travelers on this world.

You may feel free to disagree with many things in this book. This is not a research paper where I would carefully check the data before I wrote down my conclusions. For example, when I am writing about my wife's childhood I am piecing together the stories she told me, as I remember them, and she would probably disagree with at least some of it. But she and those who were there are no longer among the living. When I write about the war, I am recalling how things looked to me and what I was doing. Since I rarely knew for sure what was going on, my account reflects my ignorance. No doubt if I were to return to the places I am writing about, I would find conditions greatly changed, probably for the worse, from when I was there last. But I can't go back to Southern Rhodesia or South Africa or to Sri Lanka or Tonga. So you have been warned. This is my story.

I

# Beginnings: The Lure of Wild Country

I envy those who seem able to recall their childhood clearly. According to all accounts I was born in Mary's Help Hospital in San Francisco in 1919, at a time of great family sadness. My father had died of causes related to that year's great flu epidemic, which wiped out some twenty million others worldwide. Pictures of me as a small baby with my mother indicate that she was still doing a lot of crying. I remember nothing.

My early years were spent in a flat on 18th Street near Sanchez in San Francisco from which I made forays to Mission Park, a place then safe for families. All I remember of that dwelling was a long, dark stairway of which I was afraid. One of the building's other denizens was a man who took delight in scaring the wits out of little kids.

I have one memory of skipping along beside my mother on Dolores Street singing "a beaver, a beaver, a beaver, a bee, a beaver a company." That was my version of "vive la, vive la, vive la vie, vive la compagnie." I still don't do too well with French, but the incident may suggest an early interest in wildlife.

The only place where I really came alive during my early years was my grandparents' farm near Sonoma. I remember walking the perimeter fences with the family dog, Teddy, keeping guard. I think I remember somebody saying that Teddy had more sense than I did, which was true. I recall following behind my grandfather's horse and plow, and once finding a family of little pink and white field mice in the furrow. I remember hammering a great number of nails into a boardwalk connecting the house and barn while watching the comings and goings of barn swallows from their nests under the eaves. There were quince trees and a mulberry tree, all of them full of fruit. There was a duck pond where a gander attacked me with ferocious wing beats until I was rescued by my mother. I think we ate him for Christmas.

Then there was the rare trip to Sonoma in an honest-to-God surrey with a fringe on top. I remember the town square at that time and the candy store where I was supplied with goodies. I learned to jump up and down and yell, "De Valera" and "Erin go bragh," and I learned the words to "The Wearing of the Green" from records played on our wind-up phonograph. We were Irish all the way. The German side of my ancestry disappeared with my father's death. I don't know why. My folks were not good at explaining things or at passing on relevant information, though they talked a lot. Much later my brother Bob, after great effort and many trips to Germany, discovered that we had many German relatives, of whom none of us had heard a word.

Those were the years of the Wilson, Harding, and Coolidge presidencies. I didn't hear anything about them. They were the years of postwar prosperity, the Roaring Twenties. Prosperity passed us by and we heard no roaring. I do remember great excitement when the barn burned down. I am told that soon thereafter our horse died; its loss essentially meant the end of the family farm. I do recall the death of my grandfather when I was five, but I was unaware of what was going on and never saw him when he was ill. From sources unknown to me, my mother acquired the money to buy a block of flats on Fell Street near Webster. The building itself was old, pre-1906 (built before the earthquake and fire), with the top flat being on the luxurious side with a grand ballroom and a library. With the help of my Uncle James and my older brother Bill, the flats were converted into nine self-contained apartments. My mother undertook the work of owner/manager, renting the apartments and doing the cleaning, furnishing, and all the worrying. I made trips to the corner grocery store and later did some dishwashing, vacuuming of halls, and miscellaneous cleaning.

The house had belonged to an elderly lady named Margaret O'Callahan. When she moved to a more salubrious location on Pacific Heights, she left behind a treasure trove of stuff for us to guard, known as O'Callahan's basement. Over the years the place had filled up with furnishings, works of art, and memorabilia from the early days of California, including things belonging to General Vallejo and other famous Spanish Californians. Some in her family had traveled widely, so there were also souvenirs from far places: an ostrich egg, a carved ivory nut, wooden figurines, a derringer, a sword cane, and so on. I do believe she finally reclaimed most of it. I wish I knew more about it, but at the time I was impervious to the affairs of adults. All I remember about Miss O'Callahan is that she tried unsuccessfully to give me piano lessons, starting with bits of Beethoven's *Moonlight Sonata.* My musical education ended there.

At around the same time we moved into the Fell Street place, my aunt Margaret McDonnell, known as Peg, moved into a house in the Sunnyside District on Congo Street. Since Peg was a registered nurse and single, the rest of the family loaded her with the care of my grandmother and Aunt Nora. My grandmother at that time was seeing people who weren't there and not seeing people who were there, and my Aunt Nora was what was called feeble-minded, meaning she could not really care for herself. Peg was also saddled, part of the time, with me. Congo Street became a second home for me and from it I ventured into the semiwild highlands of San Francisco extending from Mount Davidson to Twin Peaks. There was a working dairy farm up there and lots of jackrabbits, ground squirrels, and birds. As far as I was concerned, that was the real world. The city streets were not. Our dog Teddy accompanied me at first, and later his replacement, a mixed breed coyotelike creature called Rex. In retrospect it amazes me that I encountered no people in my wanderings, and nobody worried about my poking around on my own.

In 1925 our family travels began. Bob, who was seven years older than I, suffered from asthma, and our expeditions were in search of a warm fog-free place where his health would improve. Our first base was in Los Gatos, a small town surrounded by fruit orchards. We boarded for a time at an old farmhouse owned by Mrs. Cox, along with her son Frank. I remember roaming freely through fields and orchards. Mostly I remember an attic full of World War I memorabilia, including an old gramophone that played cylindrical records with songs like "Over There" and "Yankee Doodle." There I found a beautiful first edition of *African Game Trails* by Theodore Roosevelt. It was given to me when we left, and in my irresponsible way I accomplished its virtual destruction. Nevertheless, Roosevelt's work, when I grew old enough to read and understand it, created a vision of Africa that would accompany me in my later work on that continent.

In downtown Los Gatos we later rented a house that various members of our family decreed to be haunted. This campaign was not just aimed at terrifying me—which it did—but was intended to scare the wits out of my older cousins, the daughters of Uncle James McDonnell. It succeeded in panicking them also.

Frank Cox of the Los Gatos farm was an all-around farmer-rancher who worked for a living in a place in Monterey County, in the San Antonio River valley, known as Aloha Ranch. It was owned by Charlie and Lottie Ferguson, who had moved there from Hawaii, probably during a wet cycle in California's erratic climate when the land looked green and fertile. They expected to earn a living from mixed farming, with wheat fields, dairy cows, chickens, turkeys, and a kitchen garden. The ranch house was a fascinating adobe structure with a surrounding roofed-over veranda in the Pacific Islands style. It was there that we next moved temporarily and my brother Bob did remain throughout his high school years. I went for a brief time to the grade school adjacent to Lockwood High School. To me, Aloha Ranch was paradise.

Here I first heard and later saw coyotes. At night they serenaded from a little knoll not far from the house and no doubt investigated the chicken coops. To me the chaparral and oak woodland seemed to be teeming with wildlife—quail, doves, and cottontails most noticeably, but also many hawks and a variety of songbirds. I was allowed to wander freely during our various visits. I recall discovering a remote canyon where I found the ruins of an old farmhouse and some surviving apple trees. Here I heard for the first time the mournful cry of a roadrunner. The whole canyon felt haunted. Somebody's dreams had fallen apart there in a farmstead that didn't make it.

Over the years we witnessed the steady decline of Aloha Ranch in a losing battle against drought and the competition of larger, more industrialized agriculturists. The Fergusons firmly believed that there was oil under their land, and there may have been. But it was after they died that the oil fields in nearby San Ardo were brought into production. The land now forms part of a larger land holding with access to water for irrigation. Wheat fields run across the old coyote knoll. The old adobe is in ruins.

Equally important to my life was a ranch belonging to my "rich uncle," Charlie Kaar. He made his money selling Buicks in Bakersfield and later lost a good share of it "wildcatting" for oil. The ranch was his retreat and could have served as a model for all the back-to-the-land folk of the sixties and seventies. If I had possessed common sense at the time I could have learned how to live life like that. He owned a good-sized acreage located on Walker Basin Creek below Greenhorn Mountain in the southern Sierra. On perhaps ten acres he developed a mixed-species fruit orchard—I remember peaches, cherries, and apples—using no pesticides. He brought irrigation water from the creek, which he had partially dammed, and also installed a small turbine for generating electricity. He had riding and work horses along with milk cows and beef cattle. Feed for them came from irrigated alfalfa and hay fields. He had built himself a comfortable farmhouse, barns, and various sheds. But when I went there, I was totally unaware of what he had accomplished and was only interested in running wild. My cousin Jack Gobar did not share my enthusiasm for wild country, though he often accompanied me.

Later, in my teen years, I settled on being a cowboy. Riding horses, chasing cows, and seeking wild animals were my total concern, until finally a horse threw me and kicked me in the head, but that's another story. My thoughts shortly thereafter about going into the Forest Service had no necessary connection to this event.

The Kaar Ranch with its perennial stream and extensive riparian woodlands was a wildlife paradise. Black bear and mountain lions were present, along with bobcats and many coyotes. Deer were abundant. I often sat motionless by the stream to watch deer and coyotes come to drink. I found the skull of a bighorn ram in one canyon from which these wild sheep had long since disappeared. I have not dared go back to the old Kaar Ranch. The whole area changed with the development of the Kern River and the Isabella Dam. I don't know if the old ranch survived. It does not pay to return to magical places you've left behind.

In my early years I became a voracious reader, haunting the public library and returning with arms full of books. Ernest Thompson Seton's books about wildlife had a strong influence on me, as did Will James's books about cowboy life and Jack London's *Call of the Wild.* There was another book about the fascination of learning bird identification. I no longer remember either the author or the title but it did turn me toward the most accessible wildlife in San Francisco. I set out to study birds armed with some pocket guides sold at Woolworth's concerned entirely with the eastern United States. In my first venture I identified four species of sparrows, all of which proved to be different sex and age classes of the English sparrow. Later, to my great joy, I identified a lone hermit thrush who came to abide in the garden of our neighbor's house on Fell Street.

My birding efforts took me to Golden Gate Park, where I roamed far and wide, finding juncos, white-crowned sparrows, chickadees, scrub jays and other readily identifiable species. Each trip added species to my "life list." My behavior puzzled some members of my extended family who did not share my interests. I was referred to as the "boyologist" and urged to seek more socially acceptable pursuits.

One of my greatest finds was the California Academy of Sciences, where in the North American Hall were displayed most of the species I was encountering. Also, I found there a wonderful book, *The Birds of Golden Gate Park,* by Joseph Maillard, which spurred me on to greater exploration of the park. On one occasion I met Maillard himself and was shown more of the academy's bird collection. Despite these finds, I was still a "closet ornithologist" since nobody I knew shared my interest in birds. I dreamed of the day when I could get out into the wild lands and see some of the species displayed in the California Academy.

In the 1920s the Old West was still very much alive. My brother Bill, at age seventeen, went out to Nevada to become a real cowboy on one of the last of the old-style open range cattle ranches. Later he returned to work on a cattle ranch near Tres Pinos in the inner coastal range. There he lived in an old deserted inn or roadhouse dating back to the days when travelers came on horseback or in horse-drawn carriages. He was full of tales of Joaquin Murietta and Three-fingered Jack, who were rumored to have stayed there. When we went to visit him his tales and our surroundings aroused all the old family fears of haunted houses. Pack rats in the upstairs rooms rattling their stuff around terrified my Aunt Peg in particular, but awoke us all at times. Apart from these nighttime terrors, I can recall sitting on the high seat of a cultivator watching rattlesnakes go by.

Even in those days the California farming community already had more economic importance than ranching, but places like Gilroy, King City, and San Juan Batista were basically cow towns, and livestock ranches still dominated the central coastal ranges. Most of the larger wild animals there had been hunted out of existence. Game laws were mostly unknown or ignored. Nobody enforced them. Away from the city most people hunted, but I did not know anybody with a hunting license. At the Fergusons' and later at the Kaar Ranch in my teen years I was allowed to use the available guns, and I often brought back quail, doves, and rabbits for the dinner table. However, my interest in hunting declined after my encounter with a ghost coyote.

This took place at Murphy Springs, where a pond of water remained in the dry season. Quail, doves, deer and a great variety of other wildlife came there to drink. The area had an immense attraction to me as a beginning hunter. It had the added charm of being strictly out of bounds, since it lay inside the boundary of the Hearst ranch. The vast estate of William Randolph Hearst in the Santa Lucia Mountains was reportedly patrolled by hired guns, and they were allegedly very rough on trespassers. Since I had never seen one of these fearsome riders I cannot vouch for their existence, but their possible presence added spice to the poaching of game.

One day I was sneaking around among the cottonwoods and sycamores when I saw a coyote trotting down a hill toward the water. Almost without thinking I brought my .22 rifle to my shoulder and fired. I'd gain great prestige by bringing a coyote carcass to the ranch house! Only seconds passed before a feeling of grief overtook me. I had shot my pal, the night singer from the hill. Furthermore, he/she bore a disturbing resemblance to my old dog Rex. I hastened over to where the coyote had been, feeling worse with every step. But there was no coyote, dead or alive, in the vicinity. There was no blood, no evidence that a coyote had been there. I searched the hillside. Nothing. That was the end of my trigger-happy days. I did not give up hunting then, but it took a more rational form.

The mystique of hunting is considerable. For me hunting was a rite of passage, bringing acceptance into the adult male community. The love of wild places and the thrill of stalking wary species, along with the reward of feasting off your victims, tap a mental or spiritual channel in use throughout the history of the human species. But as a way of life for all humans, hunting is no longer valid. There are too many of us and too few wild animals. Unless we decrease human numbers, hunting can only remain as a controlled and limited activity for those who pursue the sport legitimately.

A few years ago in Siberia I listened to a wildlife expert from the Baikal region tell about hunting bears in the old days. The Siberian brown bears are big, fierce creatures, kin to the grizzly. Against them hunters used the equivalent of a medieval boar spear, with a crosspiece behind the long, sharp blade to keep the wounded boar from running up the spear and goring its attacker. There are not enough bears nowadays to allow rifle hunting but—as I recommended to the Siberian expert—local authorities could make the opportunity to go bear hunting into a means of attracting tourist income to their area. The only requirement would be to use no firearms, only bear spears. Who knows, a few fearless hunters might pay high fees to engage in this suicidal activity. And the mortality rate among the bear clan could be quite bearable.

2

# School, the Woods, and War

I spent most of my childhood days not in wild country but in a place called Sacred Heart School, behind the church of the same name on Fell Street and Fillmore. Here I went with reluctance and in my first years tried to ignore my fellow students. Unfortunately, the route from my house on Fell Street to the school took me past the corner of Fell and Webster. There I came into contact with students from the local public school also on their way to morning class. Now a bunch of these public school boys lived in an alley behind our house, that is between Fell and Hayes Streets, and they were much bigger than me and mean as rattlesnakes. Perhaps because I was a small, bespectacled Catholic, they felt obliged to beat me up on any morning when our paths crossed. This involved knocking my glasses off and stepping on them along with punching me and shoving me around. Woody Allen has portrayed this whole scene in one of his movies—*Annie Hall,* I think. There was no point in trying to fight back or to run away—there were at least four of these bigger and faster kids. After a few such encounters, I learned that the rear windows of our house overlooked their houses. I could watch and see their gang get together and head off to school. Only then would I leave. I tried to get my brother Bob (seven years older then me) to come along and beat up my assailants, but he was peacefully inclined and thought I should handle my own problems.

I did put up with being hit, pushed, and otherwise threatened by schoolmates without any reprisal until I was in the eighth grade. Then an unfortunate seventh grader (bigger than me) made the mistake of shoving me. He had done this many times before. Words were exchanged. I then proceeded to beat him up thoroughly. I did not even feel a single blow. All the male students had gathered in a circle to watch this fight. For some reason the teachers did not intervene.

Following this event my status in the school changed. I became a welcomed member of the elite circle, otherwise known as a gang. Nobody messed with me anymore.

Meanwhile, back in the classroom, I became totally fascinated by history and geography, and would read the texts from cover to cover before the class had reached the second chapter. The history, I was to learn later, was biased heavily toward the Catholic viewpoint. Thus Emperor Constantine was obviously the greatest Roman emperor since he made Christianity the state religion and put down the old gods. Columbus was wonderful since he brought Catholicism to the Indians. Never mind that the Indians died.

Geography I could mull over by the hour, tracing the courses of rivers and mountain ranges. Always the blank spaces on the maps drew my attention, particularly when labeled "territory unknown." These I could envision as places where lost civilizations would still exist and miraculous beasts might dwell.

I must have been a pain to my classmates since from early on I knew the answers when no one else did. I had the catechism fairly well memorized and could make the class look good when Father Cullen, the church pastor, came down to check up on the Dominican Sisters and the state of our souls. I was appointed to the elevated status of altar boy and helped with the church masses each Sunday morning. I was clearly being groomed for higher spiritual status. My name was put forward for a special scholarship to St. Ignatius High School. Fortunately I was passed over, or I might have ended up as a Jesuit priest with results I cannot even contemplate.

By the time I had reached the eighth grade the rot had set in. My agnostic brother Bill introduced me to Hendrik Van Loon's *Story of Mankind* and H. G. Wells's *Outline of History.* These books directly challenged the safe, secure, biblical, and Christian view of humanity's origin and progress. Where were the Neanderthals and Cro-Magnon people in relation to the Garden of Eden? How about Noah and the Flood? Why were the Crusaders so corrupt and brutal, making Saladin appear a shining hero by comparison? What about the Inquisition? And so on. I became a marginal Catholic at best, observing the rituals because of family pressure. I continued with Catholic schools through the ninth grade but then was switched to Lowell High School.

In high school I developed a fanatical interest in athletics. I used to go alone to the old stadium in Golden Gate Park, where I practiced all the track and field events from shot-put to the two-mile run. I took up tennis, soccer, and, at the instigation of my brothers, fencing. I signed up for high school teams. The coach was mostly impressed by my potential as a high jumper and worked with me considerably. I cannot recall winning anything in my high school years except a ribbon in the Lake Merritt marathon. When I reached Lassen Junior College I did earn a school letter for placing in three events: high jump, high hurdles, and mile run during a tri-college meet. At San Francisco State College I was on the fencing team and a winning team in intramural soccer; I won an intramural championship in tennis. I don't know where I found the time, strength, and energy for these activities. Youth is wonderful.

At Lowell, I took a negative attitude toward classes. I mildly enjoyed a physics course in which I did well, and an economic geography class. The rest fell into an endurance category. I developed a fear of speaking up in class to the point where I would not give a verbal answer even when I knew the subject thoroughly. I now attribute this fear to the presence of members of the female gender in my classes. I had yapped away happily through grades one through nine, but in grade ten at Lowell I was suddenly tongue-tied. There were girls in my grammar school classes but they wore black uniforms and sat on the other side of the rooms. At Lowell there were many flamboyant females and some even sat at desks next to me. Fascinated by them, I dreaded making a fool of myself. I had no confidence in my masculine charm for the simple reason that I had none. I was a scrawny little kid with glasses, slow to mature and not putting on any real height growth until my senior year. So I shut up and took low grades in any course that demanded oral participation.

When I finally graduated the school counselor told me that my grades were so bad that I should not even think of going to college. This irritated me to the point where I pulled down straight A grades in my first university year and made the dean's honor list. I also forced myself to take a course in public speaking, in which I suffered sheer terror but developed some confidence. By my senior year I was a rabble-rousing orator of sorts.

During the years I was in grade school the country was supposedly glowing with postwar prosperity. The great stock market crash of 1929 did not ring in our ears. The depression that followed did register, however, as tenants became unable to pay their rent. For a time we were accepting food bags from private charity, since there was no public safety net. But I was only eleven or twelve and could scarcely tell the difference between prosperity and depression. I do remember the General Strike of 1933 when the workers shut the city down. It was fun walking down Market Street with no traffic, nobody walking or shopping. There was no question that we supported Franklin Roosevelt in 1932 and were behind the unions all the way. The New Deal brought new hope and for us, some glimmer of prosperity. My brother Bob graduated from UC Berkeley and immediately went to work as an assistant ranger in the Forest Service. Brother Bill was to follow in his footsteps two years later. My college fees and expenses were paid largely by the National Youth Authority, a New Deal agency—first as a work-study student with the Forest Service at Lassen Junior College, and later as a part-time janitor at San Francisco State University.

But while things were looking up at home, the world outside headed for trouble. If 1932 brought Roosevelt, it also brought Hitler. As the war drums began to beat again, antiwar sentiment was growing stronger in America. It was time for great antiwar movies, from *What Price Glory?* to *All Quiet on the Western Front.* I saw them all and was becoming more and more a pacifist.

With two brothers already in the Forest Service enjoying what seemed to me an ideal life in the woods, it was inevitable that I should follow in their path. At Lassen Junior College the NYA gave me a generous ten dollars a month in addition to paying my room, board, and college expenses. But with cigarettes at ten cents a pack and coffee a nickel, I did all right. We students earned our keep by "stand improvement" work, clearing brush, felling dead or diseased trees, growing young pines in the nursery and transplanting them to the woods, and sometimes doing map work in the office. When summer arrived I joined the California Division of Forestry as a firefighter. I did so well I was promoted to fire lookout at Schaeffer Mountain out in the sagebrush country. There I lived in total isolation. I had to pack water and all supplies into the lookout station. There was no electricity, and I had no visitors except deer and coyotes. I loved it.

The outside world kept interfering with my plans and career. The fall of Austria worried me. The takeover of Czechoslovakia appalled me. I was preparing to go to Lassen College when the Germans invaded Poland and World War II began. My antiwar sentiment was shaken as Hitler's tanks rolled farther and farther. When I was just learning to ski in Lassen National Park, the Soviet armies invaded Finland. I was all for joining the Finnish ski troops. But an ideological conflict was setting in. Another war, already ending in 1939, had captured my spirit—the Spanish Civil War.

Since 1936, while I was in high school, the Spanish loyalists had been fighting the fascist armies of Franco. This caused family divisions. The Catholic Church was clearly on Franco's side. My brothers and I were on the side of the republic. Most of my relatives read the *San Francisco Examiner,* which was pro-Franco. We read the *Chronicle,* which was pro-loyalist. The Russians helped the loyalists, whereas the U.S. failed to take a stand. The Nazis and Mussolini helped Franco. To me, it was all black and white (or black and red), a clear case of good versus evil. If I could have done so, I would have joined the International Brigades. If all of this suggests ideological confusion, it is true. I was confused.

I stayed on at Lassen until summer 1940, when I went to work for the Forest Service in Oregon as a lookout on Stevenson Mountain in the Ochoco National Forest. That summer of isolation was filled with strange and magical experiences. But I also suffered from being cut off from the news at a time when my known world was disintegrating. Maps I had pored over in my grammar school assured me that the sun never set on the British Empire, or the French, Dutch, or Portuguese as well. But now their hegemonies were falling apart. I decided that I must get back to San Francisco and find out what was really happening.

In the fall of 1940 I encountered a professor who rebuilt my antiwar feelings. He was Alfred Fisk, a professor of philosophy and a dedicated religious pacifist. Mahatma Gandhi was one of his heroes. His sincerity and idealism were contagious. Being twenty and full of energy, I wanted to *do* something, to fix the world *now.* I wanted to help change the political economy that seemed built on oppression and failed to acknowledge that blacks, Indians, and other minority groups had the same rights as whites. I felt that what were then colonies had the right to be free and independent from economic exploitation. I attended meetings of various radical groups. As Bob Dylan was to sing about a troubled era thirty years later, "There was music in the cafes at night and revolution in the air."

Through all this the Battle of Britain, the fall of France, and the invasion of the Soviet Union brought war closer to home. I registered for the draft but was proclaimed to be 4-F on the basis of a supposed heart condition, which in fact I did not have. That annoyed me.

In the summer of 1941 I went back to my old pursuits as a forest guard and lookout fireman at Forest Glen in the Trinity National Forest. From the guard station I could travel to various lookout stations as the occasion demanded. Forest Glen was a hamlet with a post office, a guard station, a Forest Service campground, and some cabins scattered about. When not doing more heroic things, I cleaned toilets in the campground. I was good at fire fighting and was promoted from fire fighter to crew boss to sector boss during the three summers I spent on this activity. I loved being in the forests and regarded the Forest Service as the ideal federal agency, which it may have been at that time. With the heavy exploitation of privately owned forest land there was little demand for federal timber. The situation at that time has been well portrayed in George Stewart's book *Fire* (1948).

The months rolled by until December 7, 1941. I was walking down Market Street when I saw the newspaper headlines:

Pearl Harbor bombed. I went on to a political rally where we all sang patriotic songs. I went from there to the Marine Corps to volunteer. They turned me down because of poor eyesight. I went on to the navy where I had the same experience. I went to the army. They questioned my 4-F status, put me through a thorough physical, signed me up, and told me when to report for duty.

So much for pacifism. The "bad guys" had hit my home country and were rumored to be heading for San Francisco. I was in the fight, as a private in the army. Little did I know! In retrospect, I was lucky. The marines did not ask me to fight at Guadalcanal or Iwo Jima. The navy said I could be spared from the battles of the Coral Sea and Midway. The army had some nasty surprises for me in New Guinea, but I survived.

3

# Red Arrows Never Glance

From 1941 to 1945 I wore the brown uniform of the United States Army. My rush to go out and win the war led me to sign up for the army as soon as possible after Pearl Harbor. Since I was only in my senior year in university I did not qualify to enter as an officer but had a position at the bottom of the military ladder. When the gate closed behind me in late December 1941 at the Monterey Presidio and I knew it was too late to change my mind, I began to worry that I was making a mistake. When I asked around, I discovered that none of my friends who had talked about enlisting in the armed forces had, in fact, done so. Furthermore, the reception I received at the induction center made me wonder if I had rushed to sign up for an indefinite stay in federal prison.

The main thing I remember about my stay in Monterey was being fed a Christmas dinner that was tainted in some way and led to severe diarrhea. This episode did not help me when I was put through a battery of written tests where I had to spend a good share of the time allotted sitting on the latrine, and it may have greatly reduced my IQ score. Despite that, I was assigned to combat intelligence training. This sounded pretty good. Eventually I discovered that it meant being a forward scout and observer in any future battle in which I might be involved, an extremely unhealthy position.

So from Monterey I went, via truck, to Camp Roberts, farther south in Monterey County and not very far from the Aloha Ranch where I had spent so much time in my younger years. At Camp Roberts I learned how to fight World War I, in France. We were told nothing about the ongoing World War II or about the kinds of country in which we probably would be fighting. But we could march, salute, and say "Yessir, nosir," and we knew better than to question orders. "There's a right way, a wrong way and the army way. Just do what the manual says to do." RTFM.

Leaving Camp Roberts after thirteen weeks' training, we were all moved up by night train to Fort Ord, where we would receive our permanent assignments. Hope was high among all of us that we would be assigned to posts somewhere near home. In fact we were all to be disappointed. We were all assigned to the 32d Infantry Division without regard to our backgrounds, training, or experience. Our stay at Fort Ord was brief. Soon after our arrival we again loaded on a train and traveled by night to what I believe was Fort Mason and from there onto a ship. The ship was the *Monterey,* former luxury liner of the Matson Lines, but the luxury had been removed. We were stacked on bunk beds, six to a cabin. When we pulled out after dark, passed the Golden Gate, and began to experience the deep rolling swells of the Pacific we began to realize that we were in for a long journey.

I guess all of us knew that many of our warships had been sunk in the attack on Pearl Harbor, but none of us knew that virtually the entire Pacific fleet had been put out of commission or sunk. Each day we could see our navy escort, which seemed to be a single cruiser. We were sure there must be more somewhere, just over the horizon? We kept an eye out for enemies, or rapidly approaching torpedoes.

There is only so much you can see from the deck of a liner looking out on a deep blue ocean, and you can get bored from looking at it. Before long, card decks came out and we began what turned out to be an endless game of hearts. We were told nothing about where we were going or why. But the guessing began early, at first with the thought that we might be going to hit directly at Japan. (But surely not with only thirteen weeks of basic training? Or China?) When we continued southwest and flying fish began to jump out of the deep blue, the bets shifted to India. Finally somebody discovered some large foreign copper coins in the back of a cabinet drawer. They looked like British currency but the lettering spelled Australia.

It was a cheerful bunch of GIs in early April 1942 who watched as Kangaroo Island appeared on the port side and then the harbor of Adelaide on the starboard. I was particularly pleased because the countryside looked just like central California. We disembarked to cheering crowds. They were cheering particularly because most of the Aussie divisions were in North Africa and the Japanese were at the doorstep in New Guinea. We seemed to be the main defense force, and we had never fired a shot at anyone.

The 32d Infantry Division had a long military record. We were the Red Arrow division and we even had our own battle song, "Red arrows never glance, though hell burn in advance." When we marched through the streets of Adelaide a few days later, our division band played our song and then a new song we had just adopted, "Waltzing Matilda."

Our camp was outside Adelaide near a small town called Woodside. Once we had settled in there was not too much to do. In our spare time we were free to leave the base and go into town, which had many of the characteristics of an Agatha Christie English village. Woodside had a pub, of course, its center of attraction, where we could buy Australian beer. Some of us became friends with the local minister or vicar and spent an evening at his house exchanging stories about life in America and Australia. Others took the train into Adelaide to see what could be found in a big city.

We were being trained to fight the war and therefore learned a bit about desert warfare, since it developed that we might have to hold the line against an invading force at the desert town of Alice Springs, in case of the loss of most of the Northern Territory. The northern city of Darwin was already being bombed by the Japanese and it seemed a likely site for an invasion. I was by this time part of a combat intelligence unit in the 128th infantry regiment, one of three infantry regiments in the division, the others being the 126th and 127th.

By now I had also acquired some good friends, who had been with me since Camp Roberts. Herman Bottcher was a battle-hardened veteran of the Spanish Civil War, who had known my brother Bill at San Francisco State College some years before. He was a German who had escaped before Hitler took over. He had become an American citizen and served in the Abraham Lincoln battalion of the International Brigade of the Spanish republic during the war against General Franco and the Fascist army. I had some good visits with him in Camp Roberts, but in Australia and later he was assigned to the 126th infantry and I seldom saw him.

John King was a fellow San Franciscan. He talked incessantly about his girlfriend, a *Chronicle* reporter, whom he eventually married. He and I were together throughout most of my overseas experiences.

Ozzie St. George, who came from Rochester, Minnesota, spent much of his time cartooning our overseas adventures and was to achieve fame and fortune when he put cartoons and words together in his books *C/O Postmaster* and then *Proceed Without Delay.*

One friend who was to play a critical role in my life was Bob Lonergan from Reinbeck, Iowa, who kept us entertained with Lake Wobegon–type stories about his life in a small Iowan town.

Before long we were all comfortably settled in at Camp Woodside, where the only disturbance was the Sunday church parade performed by soldiers from an Australian unit also based at Woodside. The sound of marching feet and the band playing "Onward Christian Soldiers" shattered any hopes of sleeping in on Sundays. Australia, still part of the British Empire, did not separate church and state. If you were not part of the Church of England, you were peculiar. However, we were not to be left in this relatively pleasant locale. In some way the war was moving on and we would not after all be defending the desert. Instead, we were to move to the Queensland coast to help defend the "Brisbane Line." This line was drawn by the government, who had accepted the military view that there was no hope of stopping a Japanese invasion of northern Queensland: the best plan was to make a stand just north of the capital, Brisbane. Needless to say the Queenslanders were extremely annoyed by this decision.

For some reason my friends and I were put in the advance party to establish a new camp just outside Brisbane near a place called Logan Village, not far from Tambourine Mountain. We traveled by train across southeastern Australia, mostly at night, to arrive in a woodland area where we were to construct a base camp. Here I was to have my first encounters with Australian wildlife. We had made a temporary camp in a eucalyptus woods where, while sitting around a campfire, we noticed a continued rain of pellets, some of which fell into our coffee cups. Turning our flashlights to the treetops, we saw many shining red eyes looking down at us. These belonged to the resident brush-tailed possums. Apparently they were defending their space by shitting. They were cute little critters somewhat resembling American ring-tailed cats. We covered our cups and agreed to share the territory. In the morning I wandered out from camp and encountered half a dozen wallabies, junior edition kangaroos a bit smaller than black-tailed deer, not really alarmed by our presence. However, all the clearing, noise, and confusion tied in with establishing our division's base camp eventually caused the wildlife to move on.

It surprised me at the time how quickly we adapted to new surroundings. Just as many of us could fall asleep almost anywhere whenever we had a halt in a march or other activity, we seemed to adjust rapidly to new places, settle in, and begin to enjoy whatever pleasant experiences were at hand. From our new camp we quickly made our way to the railroad into Brisbane or found rides on supply trucks or any vehicles going that way. The problem was that there were always a few who pushed the limits too far, stayed in town without permission, or otherwise failed to show up for duty. Then, of course, the weight of the military bureaucracy would descend and all our small privileges would be curtailed.

During our stay at what was to be called Camp Cable (after our first casualty in the war, a soldier killed in an attack on one of our supply ships sailing from Adelaide to Brisbane) we went out a few times to defend the coast. We took up positions on ridges overlooking potential landing sites or traveled by landing craft to offshore islands. Since no enemy was ever in sight, these were rather enjoyable excursions. They could have been more enjoyable and meaningful if somebody had explained what we were doing and why we were doing it. But that would be asking too much of the military minds. Meanwhile the war rolled on, and the most significant event took place to the north of us in New Guinea. The Japanese had pushed from Dutch New Guinea, in the west, down along the northern coast of the island, establishing bases as they moved along, in a seemingly irresistible wave. Their goals were to take over Milne Bay, at the eastern tip of the island, and Port Moresby, the capital of the Australian mandated territory in the south. These were the best harbors and the obvious launching points for the invasion of Australia.

At Milne Bay they were finally stopped, not by the proud

imperial forces tied up in North Africa, but by what the imperials called the Chockos, "chocolate soldiers," the home-guard militia. And so, moving from an entirely defensive strategy, we in the 32d were to go on the offensive and for starters take back New Guinea. We moved up to Port Moresby by ship and by air.

One of the many major and minor miracles that took place in my life occurred before the move to New Guinea. I was apparently the only enlisted man in the division with a background in meteorology. I had taken a course at San Francisco State and had also run a weather station during my three summers as a fire lookout and forest guard with the Forest Service. At division headquarters a new unit was being formed, for chemical intelligence. Captain Sandell was in charge and he picked John King and me to work for him. In case the Japanese decided to use poison gas, I was supposed to make predictions of probable weather, wind direction, and velocity in case of gas attack. Sure. In Australia! John King and I, along with Sandell and Lt. Sanders, the second in command, composed the unit.

I was still suffering from delusions about the army and the war. I seemed to think that I should hurry to the front and, when I got there, win the war and go home. So I was in the mood to volunteer for whatever would get me to confront the enemy. While still in Camp Cable I volunteered for commando training and went through an exercise or two staging mock conflicts with King, St. George, and others, in which I learned a bit about unarmed combat, karate mostly, which I have long since forgotten.

However, our leaders finally decided to acquaint us with the kind of terrain in which we might actually be fighting. So our division was sent on a 120-mile march through the rain forest. The nearest large rain-forest area from Camp Cable was the Lamington National Park, and it was there we were to experience rain-forest conditions. So we began our march. When we had reached the mountains I discovered I had an advantage over many others. Climbing mountains was old stuff for me because of my Forest Service background. I met a fellow scout and observer, Steve, who had grown up in the Rocky Mountains of Colorado. Most of the others in my regiment were flatlanders from Michigan and Wisconsin who were more used to cornfields or factories than forests. So the bulk of the regiment had difficulty climbing up the first steep slopes and considerable confusion set in. The confusion was aggravated by an encounter with a very poisonous tiger snake and there was more than a little concern about what other forms of wildlife might dwell in the forest. But Steve and I had no worries. We went forward to scout and observe. We soon found a nicely cleared trail and followed it. We had no orders to report back to anyone, so we kept on going.

As usual, we had seen no maps, nor had we any instructions about where we were going. After some miles of walking, darkness set in. So we stopped to heat up some water, make coffee, and dig into our rations. Then, refreshed, we set out again down the trail and, wonder of wonders, encountered a human settlement. It seemed reasonable to us that we should liberate it, so we did. It turned out that there was a pub and a party going on. When we went in the door with our backpacks, rifles, steel helmets, and all, we were given an enthusiastic welcome and provided with all the food and beverages we could handle. We spent a pleasant evening talking to our Aussie hosts. We found out that the place was Binna Burra, a favored stop for visitors to the national park. Eventually the rest of the regiment arrived, and the fun stopped: the officers insisted on pretending this was a serious affair.

4

# Live Coward or Dead Hero?

It became obvious eventually, though I only pieced this together much later, that our goal in November 1942 was to capture the Japanese base at the village of Buna and the nearby village of Gona on the northern coast of the island where presumably the Japanese had retreated after the defeat at Milne Bay. To do this, my former outfit, the 128th infantry, would be flown to the airstrip at Popandetta and would advance up the coast. At Buna they would meet the 126th infantry, which would have walked up a trail across the Owen Stanley Range, some peaks of which are 13,000 feet high. On the way the 126th would "mop up" any of the Japanese who were attempting to cross the range on their way to attack Port Moresby. The 127th regiment would be held in reserve to reinforce any trouble spots.

Some time after our rain-forest hike, the 128th and 126th had been moved to Port Moresby and had established a base camp there. Since King and I remained behind, we went in for various training exercises. We were asked by Sandell to participate in an experiment involving the blister gases, mustard gas, and Lewisite. It involved having a pinpoint drop of the gas placed on our arms. It was a very painful experience. In a short while we had blisters ½ inch high and the diameter of a quarter. Though these healed eventually they left scars that remained for many years. We also experienced chlorine, phosgene, and tear gases, which all in all caused us to pray we would never have to encounter them in combat. We also learned to make Molotov cocktails using gasoline and white phosphorus. I don't want to think about it.

Later, while at Port Moresby, we were assigned to an Australian unit being trained in guerrilla warfare. There we learned to handle high explosives, set and disarm booby traps, blow up bridges, and do other fun things. We actually did enjoy ourselves. The Aussies were all officers, lieutenants and captains. We had by then reached the rank of sergeants. We were the only enlisted men, but the Aussies did not discriminate. In fact, they loaned us khaki hiking shorts and short-sleeved shirts to replace our heavy combat fatigues, which were intended for colder climates than Port Moresby. All went well until one evening when we were all sitting around, telling stories and laughing a lot. Captain Sandell arrived. He said hello to the Aussies but then spotted us in our Aussie garb. He proceeded to give us hell in a loud voice for being out of uniform, impersonating officers, and so obviously having a good time. We retreated to our tents, put on our fatigues, and made ourselves scarce. The Australians, who were by then our friends, could not understand his behavior.

Captain Sandell was known among us as the Captain, though when out of earshot he was sometimes called the Whistler, because he rarely approached our tent without whistling and it was always the same tune, "Amapola" (my pretty little poppy). Sandell was a person who could inspire terror in the minds of his subordinates, though I heard one of his fellow officers refer to him as "good old Sandy." He looked a bit like Hitler's right-hand man, Joseph Goebbels.

Around the same time events occurred that made me realize I should never, never again volunteer for anything. We had discovered a civilian store in Moresby where we could buy concentrated fruit juice, canned peaches, and other goodies. I had taken the metal shell off my helmet and filled it with cold water and fruit juice bottles and hung it from a tent post. One night my tent mates and I were sitting around talking when we heard planes come over. We went out to look and then heard and saw our antiaircraft open fire. Next there was a peculiar whistling sound I had never heard before. Somebody yelled, Bombs! I rushed in to grab my helmet, spilling water and fruit juice, and frantically crammed it on over the helmet liner. We heard the bombs hit and realized we were safe, for the moment, but a lot of adrenaline had gone to work on my heart and lungs in a very few seconds. For the first time I *fully* realized that war can be really dangerous. Somebody up there had been trying to kill me! Nothing personal, of course, just because I was there.

The next event was when Sandell called us aside and told us he had arranged for us to go on a very important and dangerous mission. I believe it was the first time I had actually been shown a map. It seemed the Japanese had a base at Lae, up the coast west of Buna. The Lae harbor was full of Japanese cargo ships to supply the troops. John King and I were to be dropped off at a beach east of the town with a rubber raft and enough plastic explosive to blow a big hole below the waterline of one of these ships and sink it. What happened after that we were not told. I pointed out that I was not a good swimmer. That wouldn't matter, I was told. There were enough sharks, not to mention Japanese guns, to make that a meaningless detail.

Our proposed commando raid had gone to army corps headquarters for approval. Since I am still here, it was obviously disapproved. So much for volunteering for commando duty! We discovered that our captain was intending to make his mark and win medals, at any cost. For us, without such ambitions, it was a sobering experience.

Soon the 128th was flown over the mountains and encountered the enemy. The first man killed was the forward scout and observer, picked off by a Japanese sniper. He had taken my place on the front line. King and I had been left behind for a while, but then the word came we were to join Sandell at division headquarters. We packed up, got into a DC-3, and joined in the action. Just the two of us were passengers on that flight. We were put down on an airstrip somewhere. We did not know where we were or where to go. The pilot did not help much except to point out a road and say we should follow it. But where were the Japanese? How far away were our guys? He did not know and was in a hurry to leave. He gunned the motor and took off. We started cautiously down the road, rifles at the ready, examining every tree for snipers. It was dense jungle. We could have been surrounded by unseen enemy. Finally we reached a deserted native village and, surrounding it, the tents of our division headquarters. We asked where our captain was. Nobody seemed to know.

Lacking anything better to do, we made ourselves sleeping platforms to get off the damp ground, and after finding out where food was available, we ate and decided to turn in for the night. Little did we know we were in a tidal swamp. We were awakened when the tide came in and we found ourselves awash. We decided to move to an empty native hut built high off the ground.

The next day Sandell put in an appearance and took us up to the front line. Here I dug myself a good, deep foxhole and settled in. The next day many people I had never met decided to kill me. There was the constant whine and snap of rifle or machine-gun bullets passing nearby. A hellish, deafening roar of artillery and mortar shells made it difficult to think. All I could do was get down in my foxhole and pray I would not be hit. Then I thought that charging Japanese might be coming to bayonet me in my hole. This forced me to risk poking my head up out of my hole, hoping no sniper was waiting for me to become visible. I was petrified with fear. I might have been less scared if I could have shot back. But there was never an enemy in sight. They were concealed in the forest or in camouflaged bunkers built with coconut logs.

At some point—and I cannot recall which day—Sandell decided we should visit the perimeter of our positions and see where we were in relation to where the Japanese were. To do this he told me to follow him as he climbed up out of the trenches that had been dug by the men in one of our sectors and proceeded to stroll slowly along in plain sight of the enemy positions across a grassy flat. I marched along behind him, trying to look small, and tried to ignore our soldiers, who kept telling us to get down. I don't know why we were not killed. Perhaps the Japanese were eating lunch, or having a nap, or maybe they thought we were crazy and harmless. Whatever! Our weird parade came to an end and I was able to jump down into a trench, still alive. I realized then that Sandell was out to get a medal for bravery even if it killed him. I preferred to be a live coward.

We *were* losing a lot of men. The medics were doing a heroic job but I saw more and more body bags as the days passed. I had stopped thinking of the past and couldn't believe I had any long-term future. Living through each day was enough. *Now* was the only reality.

The war had made time meaningless to me. I lived from day to day or minute to minute, so I did not know, looking back on my experience, what year it was. And after the war our commanding general, Robert Eichelberger, wrote an article for the *Saturday Evening Post* entitled "Take Buna . . . or Don't Come Back." It confirmed my impressions. Things were even worse than I could have imagined. I had assumed that, at the worst, we had the military might of the U.S.A. behind us. We did not.

One of our major problems was the enemy's coconut log bunkers. Nothing seemed to damage their occupants. Our hand grenades bounced off them. Even the strafing runs made by our planes from time to time didn't silence their machine guns. But our air support, A-20 attack bombers I believe, certainly scared me. They put down a hail of bullets but unfortunately couldn't seem to know where we were and where the enemy was. When they came by I was really scared and dearly wished I had dug a deeper hole.

Orders from division headquarters finally hauled John and me back to work on developing some new weapons. In particular we were to see if napalm (jellied gasoline) could be adapted to flame throwers. If so, they could be good weapons against bunkers since the napalm would stick to the bunkers, continue to burn, and possibly exhaust the oxygen inside as well as covering the firing slits with flame. We got the two flame throwers that were available at headquarters and went to work. After many tests we discovered that yes, the napalm did work and did everything we had hoped for. We took the flame throwers to the front and showed a volunteer from a rifle company how to use them. He was a courageous man and may have led the first flame thrower assault in World War II. Unfortunately the flame thrower, which had worked well behind the lines, malfunctioned. Instead of being protected by a wall of flame as he ran forward, his device simply piddled out a small stream of napalm that did not come close to reaching the bunkers. He was shot, fortunately not fatally, but his heroic attack failed.

Back in the Australian guerrilla warfare training camp we had attended, King and I had been shown how to increase the damaging power of a grenade by placing it in a jam tin packed around with gelignite. We had no jam tins at Buna but we had galvanized drain pipe about 4 inches in diameter. We packed dynamite around a hand grenade in a 10-inch length of drain pipe. This could be thrown like a football for an effective distance. Using Australian TNT grenades instead of American black powder grenades, we found that we could deliver a helluva bang. Inside a bunker they could be destructive. Outside they would certainly keep the inhabitants away from their machine guns. A bit later we discovered a supply of amatol, a grayish powder explosive we could pour into the pipe around the grenade in order to get more of a bang when the grenade fired. We made up a number of these for use at the front.

Meanwhile our captain had achieved his goal. He won a purple heart, silver star, and promotion to major for his heroism. Unfortunately he was also dead. He had the bright idea of carrying two smoke pots up to a position from which the smoke would blind the Japanese in their bunkers. Under the cover of smoke, our troops could advance unseen. It may have been a good idea, but our captain decided to creep up into position himself. The path he chose turned out to be the fire lane for the machine gun in a different bunker. He did not get far before it did him in. Our reaction to the news was mixed. Yes, we were sorry, but we were also greatly relieved. Had we been with him when he had this bright idea, one of us would have carried the smoke pots.

We were left in charge of ourselves with our new officer in command, Lieutenant Sanders, back in Australia. So we carried out our flame thrower and "blast bomb" experiments without benefit of an officer's advice.

I was sitting around stuffing high explosives into drainpipes when I realized I was very ill. I was halfway unconscious before anyone noticed that I was shaking all over with chills and should go to the field hospital. The medics found that I had a fever of 105 degrees. Every nerve and muscle seemed to be hurting badly. Next thing I knew, I was on a plane headed back to our base hospital near Camp Cable with severe malaria.

Before that, however, I witnessed the end of the battle of Buna and Gona. One night I heard a noise that could only be tanks. The Australians had finally acquired them and were prepared in the morning to overrun and wipe out the bunkers, which they did. Those Japanese who were still alive moved on up the coast to our next objective at Lae and Finschhafen. We were also reinforced by the 41st division, troops who were fresh, green, and totally unaware of what lay ahead for them. What was left of the 32d Division was sent back to Camp Cable to recuperate.

I don't know how long I was in the hospital with alternating severe chills and high fever, but eventually the malaria was brought under control and I was allowed to go into camp. While walking down the road from hospital to headquarters I experienced something that made me get off the road and go sit and lean back against a fence post. Because suddenly I had not been just walking down the road but observing the world and all its people, experiencing the suffering, experiencing the joy, feeling my part in it all, and somehow knowing that everything was all right. That I need not worry about what would happen to me. If I died that was OK, but there was much I could do during my time alive on this plane of existence. I cannot really describe it: words failed me then, and more so now that many years have passed. It seemed to fit the *satori* state of Zen Buddhism or the *mokṣha* of Hinduism. During my late teen years I had tried to achieve this state in my yoga practice, but without success.

Before that moment I had truly hit bottom emotionally. My weight went from 160 pounds to 130. My friends were being scattered. Some were dead. The war seemed without end. We would go on until we were all dead. But I emerged from this experience, which may have lasted five minutes or all day, transformed. Where previously I had few friends and tended to avoid games and group activities, now I was feeling quite friendly and gregarious. I was sent to the newly established division rest camp in Queensland on the beach near Coolangatta, now part of the "gold coast" development. There we were fed excellent food to fatten us up and could amuse ourselves, free from duties. I think I actually organized a water soccer game to be played in the surf and of course, volleyball and table tennis. I was a friend to all, including to a chap named Max Ashwell, who was to be a key to the rest of my life.

Meanwhile, on another level, John King and I had written an account of our Buna experiences including details on flame throwers, napalm, our "blast bomb," and other dealings with high explosives. This was sent on by Lieutenant Sanders to division headquarters and from there to the office in the Chemical Warfare Service. This account earned us honorable mention in a return letter from the commander involved with the CWS. I was called to headquarters by the colonel in charge of ordnance who told me that though our activities were commendable, they were also suicidal. By all the rules, picking up dynamite that had been lying around in the tropical sun and stuffing it into drainpipes should have resulted in our being blown to pieces, and everything else we had done was extremely dangerous.

Even so, our activities resulted in John's being sent to Officer

Training School, from which he was to return wearing a lieutenant's bars, and in my being promoted to master sergeant. I was too "sicklied o'er with the pale cast of thought" for officer training, and besides, I was the only one left in my unit with combat experience. My friend Herman Bottcher had been promoted on the spot in Buna to captain. He had taken command of his company when the officers had been killed or wounded. Ozzie St. George had been transferred out and was a reporter for *Yank* magazine, a news magazine written by and for the army.

When I was still at Coolangatta I was informed that my turn for a furlough had come up, as had the turn for Max Ashwell. He talked me into going to Sydney with him, and there I was to meet my lifetime partner, Elizabeth Sheldon.

5

# Elizabeth's Story

Elizabeth was born in 1916 at a place called Coolum Beach near the town of Maroochydore in Queensland. At the time her family home was outside Brisbane, so readers might wonder why her mother was at Coolum Beach. The answer casts some light on her parents. It seems her father, Geoffrey Sheldon, liked to fish. Her mother, Sarah Louise Reinhold, did not fish but tended to go along with her spouse for whatever he wished to do, even though it was fairly obvious that her pregnancy was reaching its conclusion. Possibly it also provided an opportunity to get away from her mother-in-law, the formidable Gar. All might have gone well had a torrential rain not set in, flooding the only road out of Coolum Beach. So her father officiated at Elizabeth's birth, managed to tie off her umbilical cord so that it formed a neat navel, and weighed her on a fish scale at six pounds, more or less.

All this took place around midnight, so that there was some disagreement about her date of birth. Her father, who went to register her birth, put it down as September 22. He was fined by the authorities for endangering the life of a pregnant woman. I don't think he caught any fish. Louise, a more reliable witness, said her baby was born on the 23d. So Elizabeth always had the choice of being either a Virgo or a Libra and could accept whichever sign gave her the better astrological reading. In summary, she was born in a storm on a tropical beach miles from nowhere, at midnight. Obviously she was not to have an ordinary kind of life.

To begin to understand what is to follow we must dip into genealogy. On her father's side of the family were the Sheldons of Eynsham near Oxford, in England. One of them was sufficiently rich and famous to have the Sheldonian theater in Oxford named after him. No doubt he put up the money for it. Another had something to do with coal mines and the docks at Liverpool. There is supposed to be a left-wing book called *These Poor Hands* that describes how he treated his workers, or so Elizabeth was told, at a later time, by her great friend Smithy the Beachcomber, who hated the aristocracy, the gentry, and rich people in general and flew a red flag over his cabin.

With the British system of primogeniture, all the money and land went to the oldest son of the family, who was expected to take care of his relatives. Geoffrey Sheldon's line did not stem from oldest sons. But Geoffrey's father had married Frances Augusta Amelia Surtees, whose family was well-to-do and owned, at one time, much if not all of the county of Durham (for which the bull was named).

I should note that one of the Sheldons on the wealthy side of the family was chairman of the board of the Bank of New South Wales. He refused to give Elizabeth a hand when she asked for his help in finding employment in Sydney. "I know your grandmother well," he said, "but I don't know you, and without a letter from your grandmother, I cannot recommend you." A letter from that grandmother (Surtees), as you will see, was about as easy to obtain as a win in the lottery.

Elizabeth's grandfather (Sheldon) received enough money from his family to buy a plantation in Jamaica, where he went to live and grow sugarcane. All we really know about this part of his life is the scandal. He took up with a black mistress (a slave, perhaps) of whom he was quite fond. She was named RubyAnna. For some reason (the abolition of slavery?) he had to give up the Jamaica plantation and move to Australia, where he bought a cattle station (ranch) that he called the RubyAnna. Apparently Frances Augusta Amelia did not mind that sort of shenanigans. The name lives on, on Australian maps, as the RubyAnna River. As it happened, a cattle disease (pleuropneumonia) wiped out their herd, after which they bought a house and a hundred acres of land outside Brisbane.

We have a photograph showing Elizabeth's grandparents and parents in front of the bamboo grove near their house. William Sheldon cannot have lived for many years after that picture was taken. Frances Augusta Amelia, known as Gar (for grandmother), moved in with her son and his family and was to live on and on. She ruled the household with an unbending will. So far as we know, she told Geoffrey what to do and when to do it. His wife, Louise, had no say.

Louise's family, the Reinholds, were not from the upper class and were scorned by Gar as peasants. She never forgave Geoffrey for marrying Louise when he could have married money and social position. It was one time, at least, when Gar was not obeyed. Nevertheless, the Reinholds were far from being peasants. Louise's parents were accomplished musicians who met when they were performing at the Crystal Palace in London. Louise was a skilled pianist who had played many concerts before her marriage. Elizabeth remembered her piano music of an evening when friends were visiting. Mostly however, she had little time for music. Raising children was her main job, along with catering to Gar.

Elizabeth was the third child. Her elders were Cicely, who was supposed to be brilliant and talented, and Bill, the "heir" and everybody's pride and joy. Her younger siblings were born because Geoffrey wanted an "heir and a spare" in the British royal tradition. Unfortunately for Louise, she produced four girls (including Elizabeth) in succession before Gilbert, who was the last of seven, arrived. By then Louise's doctor warned her that if she had another child she would die and presented her with contraceptive materials (available only by doctor's prescription in a land where the billboards read, "Babies are our best immigrants").

Kay, the youngest, once said to us when discussing the formidable Gar, "Fortunately I escaped her attention. The old lady could only count to three." In fact, she chose *not* to count beyond two. Elizabeth existed only as her handmaiden, who cooked her food and brought it to her, combed her hair, read to her, listened to her talk, and so on. Elizabeth's mother was unable to stand up to her powerful mother-in-law and Elizabeth's siblings did little or none of the work in caring for Gar. In her later years Gar said she was afraid some of the others might poison her. She had reason to fear.

Elizabeth was to gain three things from Gar, an upper-class

British accent, a knowledge of how to behave in "high society" and on any formal occasions, and familiarity with the public library. All of these were to stand her well in days to come.

I don't know when, in the sequence of events, Geoffrey realized there was not enough money coming in to provide for his family. He was a surveyor for the Lands Department of Queensland. He took the opportunity to virtually double his salary and benefits by going to New Guinea, a hazardous area. There he became a historical figure, surveying a proposed road up the Sepik River and down the Fly River, the only north-south road in New Guinea. His work took him into country no European had visited and into contact with native tribes not inclined to be friendly toward strangers. However it meant he would be away from home for months at a time, leaving his family with Louise and Gar, not known for their friendship. To make matters worse, he arranged for his paycheck to go to his mother, so she controlled the finances while Louise did the work.

Nor do I know why or when they moved, only that they moved north of Brisbane to Gympie and then to Seaforth, on the beach just south of Mackay, in tropical Australia. Most of Elizabeth's stories relate to their Brisbane house. There they had a lot of space and were able to lease out grazing, so that at one time they had many horses staying there. The stream that ran through the land widened into a lake or lagoon in which lived fish and crayfish, and it offered lots of opportunity for kids to play in the water. Elizabeth spent much time there, catching crayfish or just sitting on a tree limb by the water, reading. There were also many fruit trees and a bamboo grove favored by snakes.

Of greatest interest to Elizabeth was the Chinese market garden. The Sheldons leased out this area of their land to an elderly Chinese who brought his relatives and others from his village in China to convert the land by the stream into a vegetable garden. They produced abundant crops and carried them into Brisbane, apparently doing very well, supporting themselves and sending money home to their village. Because of the White Australia policy intended to limit the immigration of Asians or other nonwhite races and favor immigration by northern or western Europeans, they had to go back and forth between Australia and China on limited visas and work permits. As a young child Elizabeth became very friendly with the elder Chinese, who looked after her on her many visits to his place and was also a good friend to Louise.

The Sheldon family lived in isolation from their neighbors, who were for the most part farm families. Since Gar and Geoffrey believed they were upper class, they would hardly associate with peasants. However, since they were not wealthy they could not afford private schooling for the children. I don't recall or have not heard what school Bill or Cicely attended, but I know that Elizabeth went to the local country school. Here she had no friends and was a target for the local louts, who at times threw rocks at her. To add to the problem, she was an avid reader and had read all the books in her house, including Gibbon's *Decline and Fall of the Roman Empire,* Macaulay's *Lays of Ancient Rome,* the poetry of Tennyson, Byron, and Browning, Shakespeare, and a book labeled Euclid.

The only time she really won some appreciation from her schoolmates was when an inspector from the education department came to check on how well her teacher was teaching and the students were learning. The teacher, wishing to make a good impression, called first on Elizabeth to recite a poem. She rose to her feet, shut her eyes, and began with, "Lars Porsena of Clusium / by the nine gods he swore / that the great house of Tarquin / should suffer wrong no more" and carried on without pausing through all umpteen verses of her favorite poem, "Horatius at the Bridge." The inspector decided that he had heard enough and allowed the class to return to its normal torpor, to the great relief of students and teacher alike.

She did get back at the stone throwers when she caught them smashing swallows' nests under the bridge across the local stream. She was in a native canoe that her father had brought from New Guinea and, after yelling at them to stop, she charged their rowboat, swinging her paddle at them. She forced them into the water and left them to swim or sink, she didn't care which, and went on downstream to home.

The native canoe featured in another story when she was out poling the canoe around the lagoon gathering yellow waterlilies. An American car happened to come along with a tourist couple. The woman passenger told the man driving, "Stop! Stop! There's a native!" She rushed out of the car and took photographs of Elizabeth and her canoe. No doubt somewhere in America her pictures have been shown and she has been pointed out as a genuine Australian aborigine.

Elizabeth's scholastic abilities caused her to be promoted ahead of her class and led to her being sent at age twelve to high school in Brisbane. There she had a really bad time, since she was immature for her age, small and undeveloped, and found herself surrounded by girls who were past puberty and well into adolescence. Once again she could not fit in and also was up against those from the Brisbane city schools who had already received teaching in subjects that her country schoolteacher could not cover. Nevertheless she somehow survived, graduated, and was able to go on to study art, her primary interest in life.

At that time in Australia, art was taught in technical college, not in university. She enrolled therefore at the Queensland Technical College in Brisbane, where she was able to devote her time to drawing, painting, and, to a lesser extent, other areas of art. She also made friends among the students and could broaden her knowledge of the world beyond what her family provided. Her great goal at that time was to get away from her family and get to Sydney, which was more of a center for art. However, she had been led to believe that she could not leave home and find work on her own until age twenty-one. In fact, she could have left at eighteen, but this she did not know. She did know, however, that the excuse of early morning and evening classes allowed her to leave home, catch a morning train into Brisbane and a return train in the evening, and thus avoid much of the strain of dealing with Gar and her siblings.

Elizabeth's parents, both together and separately, devoted much time to the upbringing of Bill and Cicely. In this they were joined by Gar. That left relatively little time for Elizabeth, Marjorie, Dora, and Kay. Marjorie in particular was an object of Gar's dislike. She was blonde, blue-eyed, and chubby and looked very much as though she had inherited a full array of Swiss genes from her mother's ancestors. Gar refused to have her in the same room and proclaimed that she was not a member of the family at all, thus indicating that Louise had strayed outside the boundaries. Louise, of course, had no time to stray anywhere and was always up to her neck in child care, pregnancy, or hospitalization resulting from these excessive activities.

Cicely, supposed to be a brilliant scholar, was allowed to devote her time to her studies, and in fact she did well in school and went on to university for a degree in agriculture. However, she refused to take any job that would require her to leave home; so as far as I know, she never did work for a living. This fitted in well with her father's idea that upper-class women did not work for pay.

Bill was expected to bear all the family hopes. However, efforts to educate him generally failed. He was interested in physical activities and not in books. Unfortunately, in an effort to spur him on they put him in competition with Elizabeth. The poor kid flunked, while she excelled. To make matters worse, Geoffrey decided Bill had to be taught the "manly arts," and boxing gloves were produced. Following the usual pattern, Elizabeth was designated as his punching bag. However, both father and son learned an important lesson. When Elizabeth became angry, she was able to draw on a "berserker" form of energy and strength. Never give up, never surrender. Poor Bill decided not to fight with her any more.

When Elizabeth was five or six years old she had an incessant cough and became quite ill. Her parents feared that she might have tuberculosis. The fact of her illness did not seem to worry them much, but the thought that Bill might catch TB from her scared them. They decided to send her away to stay with Jenny, of whom she was quite fond. Jenny was a Solomon Island native and was one of a pair of Solomon Island girls the

Sheldons had acquired as servants—don't ask how since I do not know. Because Jenny had married an Australian, she was allowed to stay in Australia when the White Australia policy was enforced and her sister was deported to the Solomons. So Elizabeth went to stay with Jenny, a great improvement over home. Unfortunately Jenny became ill and could not care for her, so Elizabeth went back to the family home. Not long afterwards, Jenny died of tuberculosis. Elizabeth, despite her association with Jenny, did not have TB. What was causing her cough was only discovered when a medical team from the Rockefeller Foundation came through on a survey of the distribution of tropical diseases in Australia. They discovered that Elizabeth had hookworm, apparently picked up from her association with the Chinese market gardeners, since she was the only Australian child in her area with that infection. The appropriate medicine quickly cured it. Her cough disappeared, but she retained a tendency to cough under stress, which was often. She also developed a tendency to hide from her family and established play areas under the front veranda, up in trees, or anywhere she would not be easily discovered. This behavior led various family members to consider her sly and sneaky. Sitting in her hiding place, Elizabeth once overheard her mother's Auntie Dill say, "You have to watch that one, Louise. Still waters run deep."

# 6

# Reunion

I did not recall much about the sequence of events in Elizabeth's move to Sydney. Evidently when she first brought it up, Gar decided that she herself should go along to introduce Elizabeth to the right people. There were various relatives in Sydney, so they could have picked her up. But this was what Elizabeth wanted to avoid. She managed to shake herself loose and be on her own. I know that she worked for one company on fabric design. She also returned home after the first visit and worked with her father in his survey camp. At some point she was employed as a librarian on the *Sydney Daily Telegraph* newspaper and thus was exempted from the armed forces women's auxiliary (WAACs), since newspapers were considered essential services during the war. She spent a lot of time with the Black and White Artists Club and developed a circle of friends, mostly radicals, artists, writers, and journalists like those I associated with then at San Francisco State, half a world away.

I think it was on her second trip to Sydney that she had an

unusual vision while dozing on the train. In her dreamy state, she saw a room with a sign on the door reading Danger, Keep Out, Fumigation in Process. When she arrived and was looking for a place to stay, she encountered a friend from the artists' club who told her a studio apartment was becoming available in the building where he and his friends lived. When she met the manager and agreed to rent the studio, she was told that she was lucky, it was the only apartment available, and that until yesterday it had had a quarantine sign on the door. So fate put her at 67 Macleay Street, where she became good friends with Norma Norton, an artist for the Sun newspapers. During a vacation spent on the beach at Coolangatta, Norma had become acquainted with Max Ashwell, who talked me into going to Sydney with him for our furlough.

So it was all arranged for us by our joint fates. And it had been foreseen. When Elizabeth was sixteen years old she listened to a "bush psychic" said to have the ability to see people's futures. She was sitting with a group of girlfriends around a campfire where he had been asked to tell each of them when they would marry and whom. As he predicted each girl's fate, there was a lot of laughter and joking. When he got to Elizabeth, he told her she would not marry until she was twenty-seven and then would marry a man who came on a ship from overseas. She thought his words were nonsense and buried them in the back of her mind until it dawned on her in 1943 that she had fulfilled the prophecy.

Norma had persuaded Elizabeth to take a chance on a blind date with a Yank soldier, and Max arranged a double date with Norma and her friend for the evening following our arrival in Sydney. Max and I arrived at Norma's place, and I sat down on one of her couches with my head against a cushion against the wall. Then Elizabeth arrived and I felt enormously impressed and relieved. Not only was she beautiful, but she seemed to be someone I already knew. We sat around talking, joking, drinking a bit of beer as though we were old friends. Elizabeth shared my cushion. We were already close. When finally she had to leave, since her work started early in the morning, I went with her. I felt I must see her again soon and we agreed to meet for lunch the next day. I had no idea at the time how she felt about me, but I was to learn that our feelings were mutual.

I certainly lack total recall of the events that followed. We had meals together and once made a raid on a delicatessen for treats like cheese and salami. We went dancing at a nightclub, along with Norma and Max. We visited the home of her old friends Neil and Gwenda Moody, and I met their toddler, a girl named Robin, of whom Elizabeth was very fond. It was like a dream world for me. The war and the army seemed far away. Neil was a journalist for the *Telegraph* and wrote a totally fictitious account of our experience while walking from the suburb of Cremorne to Mossman, where the Moodys lived. It was published the next day as a news item.

At last reality closed in on us: my furlough time expired and I had to leave. Physically I went back to Camp Cable. Mentally and emotionally I never left Sydney. After I had been moping and whining around camp for a week or so, spending my time writing letters to Elizabeth, Bob Lonergan said, "We lost a good soldier when you went to Sydney. He never came back." When I received a letter from Elizabeth, I realized that she shared my feelings. I was determined to get back to Sydney even though we knew that the 32d was moving its base to New Guinea.

Another in a series of small miracles then took place. Bob Lonergan's turn for a furlough came up, but he had a girlfriend in Brisbane and did not want to go anywhere. He turned his furlough over to me. This sort of thing never happens. Furloughs were rare and priceless for us GIs. I was and am eternally grateful to Lonergan.

So we went home to Elizabeth's apartment. On September 29, 1943, that same day, we decided to get married as soon as possible. We explored the mysteries of sex together and decided we liked it. We would have married even if we were both sworn to celibacy. We just wanted to be together. Always.

But we could not just rush into marriage. It had to be approved by all the official channels if we were to get permission for her to come to America or for me to stay in Australia. Meanwhile we had three wonderful weeks where we were engaged and living happily in sin. She stayed with her job on the *Telegraph* but we had evenings, weekends, and some days off when she could manage to get free.

Then came the horrible experience of parting again. On the day my furlough expired, I was told to check in at the base at 8:00 A.M. and be ready to board ship. But then after waiting around for hours we were told our departure was postponed and I should come the next morning. This same routine was to go on for a week. Each morning a sad good-bye, not knowing how we could ever manage to meet again, and then each afternoon, reprieve.

Finally the day came when I couldn't go home. The journey out was sheer torture and included seventeen days in the harbor at Townsville, where I slept on deck, on a board, and had a bout of what was called dengue fever, though it may have been simply a recurrence of malaria.

Finally we landed at a base already established at Finschhafen. There I had an interesting time exploring a coral reef and writing an endless stream of letters, including one to the commanding general requesting permission to marry an Australian woman.

We were not to stay in Finschhafen for long. Soon we packed up and moved to Goodenough Island, one of the D'Entrecasteaux Islands off the northeastern coast of New Guinea. Why we went there I did not know. Our commanding officer evidently thought we would be there a long while, since he arranged to plow up a kunai grass flat and plant it to vegetables. The island featured a 10,000-foot volcano I wanted to climb just to get away from the oppressive heat and humidity on the flats where our camp was located. We were told not to try it. The higher elevations had hostile natives, along with some Japanese soldiers.

It was there I encountered what I was told were giant stick insects. They were up to 9 or 10 inches long, ½ inch across, and armor plated. I did not know what they were, only that they looked dangerous and that a whole platoon of them had moved onto my bed. I proceeded to attack them with a machete, strewing insect parts about the tent. Our camp also supported giant black spiders the size of a dinner plate, and huge red centipedes, as well as the usual squadrons of mosquitoes.

On Goodenough Island, all of us were a bit edgy about possible night attacks by the Japanese or unfriendly natives who were still on the island. We had an established camp with the big pyramidal tents that accommodated four army cots along with our equipment, but we did not feel really secure. Furthermore, in both tropical Australia and New Guinea we had encountered a fair number of big pythons. These constrictors were known affectionately as carpet snakes by rural Australians, who welcomed their presence in barns and farm outbuildings to help keep the rat population under control. They were inclined to grow to 12 feet or more in length and were quite formidable in appearance.

One night when all of us were asleep, I dreamed I was being attacked by a large python that was crawling across my chest. In the dream, I grabbed it by the neck and was pulling it off me. As I struggled, I overturned my cot and landed on the ground, partly wrapped up in my mosquito netting. My loud yells scared the wits out of my tent mates. When they found their guns and flashlights, they discovered me yelling and fighting on the ground. I had grabbed my left arm with my right arm and was trying to beat it to death. What had happened was that my left arm had "gone to sleep," becoming numb and losing all feeling. As I turned over in my sleep, my left arm dropped onto my chest, coinciding with the python attack in my dream. My tent mates, who had already put up with my attack on the giant walking sticks, considered asking me to move to another tent. But they hesitated: in another tent, with less tolerant tent mates, what if I touched off an armed conflict?

We were well settled in on Goodenough when we got orders to move once again. This time it was to Aitape, where I believe our purpose was to stop the Japanese who were retreating up the coast. However, it was here that what seemed like another miracle happened. In response to my request, I received a letter from the commanding general giving me permission to marry Elizabeth and granting me another three weeks' leave, to begin upon my arrival in Sydney. By that time, eight months after our unofficial marriage, I would have walked or swum to get back to Elizabeth. No transportation would be provided. It was up to me to find my way back to Sydney.

Fortunately I was not alone. The 32d Division band had been granted leave, along with a lieutenant from the medical corps. We formed ourselves into an unofficial detachment with the lieutenant in command, me as first sergeant, and the band members as the enlisted men. In this guise we hitched a ride on a transport plane over the mountains to an interior base at Nadzab. From there we hitched a ride on trucks going to Lae. In Lae we talked our way onto a Dutch transport ship heading to Milne Bay. There, however, we had to wait for a ship heading to Sydney. So the band brought out their instruments and for several nights serenaded all the troops based at Milne Bay.

On the trip aboard the Dutch ship from Lae to Milne Bay, I saw what appeared to be an ideal tropical island. It was not visibly touched by war. Too small to be worth capturing or recapturing by the military, it stood in isolation from the forces of destruction and chaos battling back and forth on land. There was a white British colonial-style house with wide verandas and associated sheds and outbuildings in a grove of coconut palms with a backdrop of rain forest. It seemed an ideal place for a family to live and reinforced my fascination with islands. Admittedly, New Guinea itself is an island, but in my way of thinking it is much too large to qualify for island status.

I wanted to come back to this perfect island when the war was over. But fate decreed otherwise. I don't even know its name. Call it Bali H'ai.

Always in New Guinea we were surrounded by tropical forest. It varied from the mangrove groves along the coast to the mature rain forest higher up, to the wet and cold cloud forests of the mountains. All of it included both mature forest with tall trees, often buttressed, and also developmental stages where the mature forest had been destroyed and dense undergrowth had developed. These areas were truly jungle and could be almost impenetrable.

Even more formidable than the jungle were the kunai grasslands. Not only did the kunai grow tall and dense but the grass had serrated edges that could slash you if you tried to force your way through the stands of grass. The air temperature and humidity were much higher in the grasslands than in the forest, and there would be little air movement even when breezes were blowing outside the kunai fields.

I was in no condition to learn more about this tropical paradise. Nobody I knew had any knowledge of tropical forests and there were no books available where I was, nor even the leisure for exploration. Survival was on all our minds as a first priority, and secondarily, at best, winning the war. There was little to be seen in the way of wildlife, though New Guinea has an amazingly diverse fauna and a particularly rich bird population. But then, no sensible animal would stay where we were.

The trip from Milne Bay to Sydney was a rough one. May is the beginning of winter in Australia and we certainly encountered a winter storm. Once, when we were all having breakfast in the mess hall, we hit a swell that sent all our aluminum trays right off the table and into the wall and then sent them all flying toward the opposite wall. Fortunately I was not inclined to seasickness and had a firm belief that, having gone through as much as we had, we would not sink. We didn't.

To say that Elizabeth and I were ecstatic about our second reunion would be to state the obvious. Since we wanted official marital status as soon as possible, we arranged to get married at the first available time at the registry office in downtown Sydney. This was May 30, 1944. We asked my friend from Camp Roberts, Ozzie St. George, who was now a writer for *Yank* magazine and based in Sydney, to be one of the two witnesses required for the wedding. We asked Norma to be the other. But any hope we could have a quiet wedding vanished when we arrived at the registry office.

All of Elizabeth's friends were there and most of them had bottles of alcoholic beverages concealed about their persons. A car with Just Married signs had been arranged, and after the moving and very serious ceremony we were transported across the harbor bridge to somebody's house in the suburbs. The party was fun and went on into the night, but our friends finally released us to the quiet of our own apartment. We had a few weeks of happy domesticity interrupted by parties, dances, movies. It was a beautiful time in our lives.

I arranged for Elizabeth to be the primary beneficiary on my life insurance and the recipient of all the money I had accrued from the army. We decided it would be a better thing for us to go to America than for me to stay in Australia, and we were to concentrate on doing whatever was necessary to get us both safely across the Pacific. For her, this meant a series of examinations: medical (yes, she was healthy), psychiatric (no, she was not insane), and so on. Finally they decided she was fit to be an army wife and an American citizen, and she was told she would be taken to San Francisco when a ship was available. She left her job on the *Telegraph* in favor of a commercial art job and later, when it seemed that the likelihood of her shipping out was more imminent, she went to work for the British Navy. She had to be ready to leave on short notice.

For me the principal duty was to stay alive. During our landings at Finschhafen and Goodenough Island there had been no serious fighting that I heard of. Of course, headquarters was far in the rear and I did not really know what was going on. At Aitape it was different. After I had left for Sydney in May, a major fight developed along the Drinumor River. I think the Japanese were using guerrilla tactics by then.

When I arrived back from Sydney and found my way to headquarters, it turned out that I was a day late for the excitement. One squad or platoon of Japanese had waded along through the surf and had come in on the beach to attack. Since our defenses were aimed against an attack from inland and not from the ocean, and the attack took place during the night, our troops were unprepared. I heard that Colonel Smith was bayoneted in the butt as he was getting into a tank.

Fortunately our men got it together sufficiently to wipe out the attackers, but when I arrived they were still badly shaken. And as I put up my jungle hammock and crawled in to sleep, the prospect of a bayonet in the dark really got to me. I pictured an enemy creeping up out of the swamp, bayonet at the ready, while I was zipped into my mosquito-netting-lined hammock like a Christmas turkey swinging in the air. There were enough animal noises coming from the swamp to sound like a full company of enemy soldiers. I abandoned the hammock and joined my comrades sleeping on the ground. To hell with mosquitoes. I awoke in the morning unharmed.

Somehow or other we won on the Drinumor. The next move was to Hollandia (now Jayapura), the capital of Dutch New Guinea. I waded in bravely enough through the waves at Hollandia but the battle, if any, was long since over. We set up a full-scale headquarters camp there while we wondered where we would go next. The battles for New Guinea were over and some of us old-timers were still alive.

Up to this point we had been paid each month in Australian pounds, shillings, and pence. Now, however, we were paid in Dutch guilders. I did not see any Dutch people around, but it may be there were some. Dutch New Guinea is now Irian Jaya and part of Indonesia. Japanese have been replaced by Javanese. Either way the natives have suffered.

The big news that reached us in Hollandia was that we would be sent home. A point system had been set up, so many points for length of time spent overseas, more for being in a combat zone, and so on. I had a lot of points and before too long I qualified. Came the time to board a reconverted freighter and head out across the Pacific. To do this I went out to the ship on a landing craft and then had to transfer to a rope ladder to climb on board. Unfortunately the landing craft was bobbing up and down and slamming into the side of the ship. The man in front of me got hold of the ladder but then froze in position. The landing craft rose up and slammed into his legs. He had to be rescued and taken to the base hospital. With his example in mind, I went up the ladder like a scared monkey.

Of course the war was not over and there were submarines

out there. We hoped the navy was doing its job. It turned out we were on a really bad ship. Not only were we overcrowded and the chow lines interminable, but the food was terrible. Mostly it seemed to be rice that we had captured from the enemy and that had been left around too long in the New Guinea sun and rain. Sometimes we had mystery meat we suspected might be rat. But what the hell, we were going home!

The trouble was I didn't want to be going home; for me, home had become Sydney, but the army was not shipping us to Australia. I had no clue where Elizabeth was, since mail service during the war was usually delayed for censorship and slow at best. I was happy to finally sail under the Golden Gate Bridge, but I dearly hoped that Elizabeth would be there before me, staying with my family and waiting for my arrival. She was not. My homecoming was a mixed affair. I was glad to see my folks after nearly four years but really worried about my wife running the submarine gauntlet.

Then, to top it off, I was given a newspaper clipping about the death of Herman Bottcher, a captain by then, and leading a diversionary assault on the island of Halmahera. Presumably it was intended to make the enemy think that the 32d Division was going there instead of to Leyte in the Philippines. It was a suicide mission. For the first time in the war, I broke down and cried.

7

# Transition

My leave period in San Francisco involved visits with all my relatives. These were cheerful enough occasions, but nobody seemed to have the slightest interest in where I had been or what I had been doing. I was treated as though I had just returned from a four-year vacation in Mexico, except that nobody asked any questions about life there. Later I found out that they had heard via radio or newspaper that they should not ask returning veterans about their war experiences, because it would be too upsetting to us. It was certainly disturbing to have no interest displayed, when we thought we were out there saving the free world.

When my home leave ran out, I was taken to Santa Barbara, where the military had taken over all the luxury beach hotels for our use. It was a great change from our previous quarters and meals, but I was missing Elizabeth and nobody could cheer me up. I was really upset to learn that the army was planning on sending me back across the Pacific, not to

Australia but closer to Japan, to join in once again in the war. However, luck was on my side once more, or perhaps it was that long-forgotten course in meteorology. I was transferred to the air corps and shipped to Wichita Falls, Texas. It was a dismal place but better than Iwo Jima. I had barely settled down in Texas when I was moved to Lowry Field in Denver. There they were once more threatening to send me overseas. Oddly enough, they had decided I would make a good B-29 mechanic and could work on our big bombers out in Okinawa. But a recurrence of malaria put me in the hospital again. It was fortunate not only for me but for anyone who had to fly a plane I might work on. I had the mechanical ability of a white rat.

One day when I went into Denver to look around I shared a taxi with a lieutenant from the chemical warfare section at Buckley Field. He told me they might be able to use me at Buckley Field and thought I could apply for a transfer. This seemed a way to get off the list of those awaiting assignment, since most assignments seemed awfully close to Japan. So I trotted over to Buckley Field the next day and went to see the colonel in charge of the chemical warfare section. He gave me hell. Asked if I had not heard of such a thing as proper channels. I told him things were more informal in New Guinea and did a lot of yes-sir, no-sirs to him while the lieutenant I had talked to tried to look invisible across the room. Finally after much interrogation he agreed to request my transfer to his section. So for the rest of the war I lectured to classes and demonstrated how to deal with mustard gas and other nasty stuff.

In the midst of all that I received the news that Elizabeth

was on her way. My mother and Aunt Peg had cooked up a story about the need for me to meet Elizabeth when she arrived and had worked through the Red Cross to arrange an emergency leave for me. Suddenly I was on the Greyhound bus to San Francisco and another reunion at long last. Hopefully a reunion not to be interrupted again by wars or armies of any kind.

Elizabeth came across the Pacific on the *Lurline,* the Matson Line sister ship to the *Monterey,* which had taken me to Australia. It had been converted to a "war bride" ship and was loaded to the gunwales with women who had married Americans. There was, of course, more than a little resentment against the Yanks running off with Australia's most beautiful women. It was widely predicted that all such wartime marriages were bound to end in divorce, and that anyway the women who married Americans were just in it for the money. Elizabeth's Auntie Dill pontificated that Elizabeth had just married for adventure. I do believe she found that, but not exactly the sort Dill meant.

We settled in at Denver. Apartments were scarce because of the number of military families stationed there, but we did find a "studio" apartment: one room and a closet converted to a kitchenette. We were just happy to be there. I believe that it was during my "emergency" leave that we visited my older brother Bill, who was a range manager with the Forest Service in Globe, Arizona. We thought we could do pretty well with the kind of house the Forest Service provided for him. However, they had a small child, Gary, and in consequence Bill's wife, Edna, worried constantly about scorpions, centipedes, rattlesnakes, and such. They finally had to give up the

Forest Service and move back to California, where he joined the Department of Fish and Game.

While we were visiting, we heard how they had wiped out Hiroshima and then Nagasaki with atomic bombs. Even though I would not want to fight the Japanese on their home turf, we thought that was an overly drastic way to win a war. We went from Arizona to San Francisco and arrived just when everybody went crazy having heard that the war was over. We managed to escape the crowd and get back to the family home. I did look forward with the greatest of pleasure to getting out of the damned army. It did not take too long before I received my discharge papers and was a free man.

Our apartment in Denver had one of those fold-down beds. When it was down the bed occupied most of the room, so there was not much we could do except go to bed. And fairly soon we found out that Elizabeth was pregnant. She had some complications that resulted in bleeding. The doctor recommended that we go by train to San Francisco with a compartment where she could lie down for most of the journey. We did that and before very much time had passed she was having a normal pregnancy and was full of her usual energy and happiness. We found a nice apartment on California Street in the Nob Hill area and started to enjoy life in San Francisco. My friend John King and his wife, Edna, also arrived in San Francisco, so we went out to movies, plays, ballets, and other events with them often.

We both worked at miscellaneous jobs. I was a civilian engineering draftsman with the navy at Hunter's Point. But I wasn't a good or even a fair draftsman, so in due time it was suggested that I resign before I was fired. Just as well. Some of the stuff that I drew and made blueprints from could well have sunk ships. One fine day we had a phone call from an old friend of Elizabeth's, Annette Kellerman, who was a professional swimmer. I believe she swam the English Channel, along with other feats. She lived on an island offshore from the Sheldon family place at Seaforth. We had to dash down to the Embarcadero for a brief reunion with her before she caught a ferry and then a train for New York. So we jumped on a cable car and managed to have coffee and conversation. As we were getting on a cable to go back up California Street, the imminent arrival of a child made itself urgently apparent. We went to our apartment to grab some of Elizabeth's stuff and our car and proceeded to the hospital.

There began one of many adverse experiences with the medical profession. Our doctor was supposed to be one of the best specialists in the area, but he had decided that Elizabeth was not due for another week and had gone golfing. The intern who was available was fresh out of the navy and knew little about childbirth, but he and the nurses decided that she would not go into real labor until the next day. So I was told to go home and come back in the morning. As a result, Elizabeth could not get any attention until Sandra was well on the way and Elizabeth had serious tearing. The great specialist arrived that day (April 26, 1946) in time to sew her up. She was kept in the hospital for ten days but not told any details about the birth, and she did not find out about the injury until we encountered a reliable doctor some years later. However, Sandra was not harmed. And we thought (of course) that she was the most beautiful child on earth. Elizabeth managed to bounce back from her experience.

Meanwhile, we had reached a turning point in the life of my family. My mother had to get out of the apartment business and in May 1946 she put the Fell Street house up for sale. Since we were all very naive when it came to real estate values, she sold it too soon for too little. Elizabeth and I found that our apartment, while a great place for two of us, was overcrowded for three. So we decided to join forces with my mother. She would buy the house, but we would make the mortgage payments, and we would gain a part-time babysitter in the bargain.

I was sort of drifting at the time with no clear goal in mind except that of getting Elizabeth back into the art business. She had the goal of getting me to finish my university education so that we could get a Forest Service or similar job and live in the woods. She prevailed. With the help of the GI Bill of Rights I signed up for forestry at UC Berkeley. She went to work for the *San Francisco News,* and later the *Oakland Tribune*, as a librarian. I received a small stipend under the GI bill and also, because of my malaria, a disability pension. She received a full salary.

We bought a house on Liberty Street for which we paid $8,000. The same house is often shown in photographs today when people want to tell what a wonderful place is San Francisco. I imagine it would sell for a million. It was built on the side of a hill with three livable floors and storage space underneath. It had a view of downtown and the bay. At that time, however, it had only one bathroom, and since my mother could not navigate stairs very well, only the ground floor was really available to her. We ended up selling it for $9,000 and buying a one-story house on Sonoma Avenue in Berkeley, where we remained for the duration of my undergraduate years at UC.

When I started at the university I found that UC did not accept a number of the courses I had taken at Lassen Junior College and it was necessary to repeat several of them. In dendrology, for example, we used the same textbook, but I have to say I learned more at Lassen than I did at Berkeley. The course at both places dealt with tree identification with emphasis on timber trees that could be converted into "useful" products. As time went on it became obvious that the emphasis in the forestry program would be on the practical, lumber production of forests rather than on their value as part of a biotic community that had intrinsic worth standing uncut, little disturbed, in a wild landscape.

There was, however, a course on plant ecology that awakened my interest in the field of ecology. The textbook *Plant Ecology* (1938) by Weaver and Clements acquainted me with the Clementsian theory of plant succession, from pioneer plants establishing themselves on bare ground through various developmental stages to a climax formation of trees, shrubs, and herbaceous plants able to hold the ground permanently, fully adjusted to the region's climate. I found Clements's theory very intellectually satisfying, along with its classification of the world's plant communities into climax formations appropriate to each regional climate. Thus in North America, his theory predicted that a mosaic of developmental stages—marshes and meadows, aspens or birch—would, in time, develop into spruce/fir. It was not until later in my reading that I learned there were many problems with Clements's theory, including the problems associated with frequent, severe disturbance.

What finally caused me to drop a forestry emphasis in my university program was not a lumber production/biotic community issue. It was the fact that those in charge of the program treated the postwar generation of students like the eighteen-year-old, fresh-from-high-school students of earlier years. Many of us were married with children, working part- or full-time in addition to doing university courses. The idea of a six-week summer camp requirement might have been appropriate for those unacquainted with survival in the out-of-doors. No, we could not bring our families with us. So what were those of us from the battles of Normandy or Anzio, New Guinea or Iwo Jima supposed to do? Get divorced so we could camp in the woods?

When I received the news that Starker Leopold, son of Aldo Leopold, was starting a program in wildlife management in the zoology department, I hurried over to zoology to apply. There was no problem and I would be able to take courses of greater interest to me than those offered by forestry. In my new program was an ecology course taught by Frank Pitelka, a range management course from Arthur Sampson, and one in conservation of natural resources taught by Carl Sauer, along with the three courses taught by Leopold. In fact, since Starker was on leave for one semester, one of his courses was taught by Carl Koford, who was to become the expert on condors.

Without having a particular goal in mind, I completed courses that led me into an interdisciplinary approach to research and other work that I was to use throughout my career.

Pitelka's ecology course was far broader in scope than the plant ecology I had studied in forestry and brought me in touch with the ideas of such animal ecologists as Victor Shelford—who built from the Clementsian plant formations to develop the biome system for classifying the earth's major ecological communities (such as tropical rain forests or grasslands or deserts)—and the British animal ecologists Charles Elton and Frank Fraser Darling. Carl Sauer, in the geography department, was the ultimate interdisciplinarian, drawing on history, archaeology, and anthropology as well as the natural sciences to present a view on how humans had used and abused the resources provided by nature. Having also picked up courses at San Francisco State in political science, economics, and sociology along with geology and biology, I was equipped to go wandering through interdisciplinary fields. Of course each of these fields had its staunch defenders, who gained fame or promotion for their work in specialized parts of their disciplines and rarely appreciated generalists who tried to combine their specialties with those of others who spoke in different terms. But I did not realize the dangers of trespassing on the intellectual turf of specialists until much later.

On the home front, meanwhile, all was not well. I had unwittingly asked Elizabeth to move back into a situation resembling in many respects what she had escaped from when she left home. Though my mother did not equal Gar as a tyrant, the two had enough points in common to trigger the same emotional response in Elizabeth. The psychological wear and tear was rough on both of us, but mostly on her. I did not know what to do about it. We had barely enough money to live on, let alone split up into two households.

Fortunately I did graduate and—thanks to Starker Leopold—had a paid job with the opportunity to do a master's degree thesis as part of a research project on the California deer populations. We moved off to a cabin in the woods near Sonora, while my mother sold the house and moved to a flat in San Francisco's Richmond District.

8

# Deer

I was delighted to have a paying job that would also lead to a master's thesis and degree. But I don't think that in 1948 I fully appreciated the opportunity Starker Leopold had given me. The work was to be part of an overall deer study supported by the California Department of Fish and Game and contracted out to UC. The department set out to determine the actual status of California's deer populations and the condition of their habitat, in an attempt to deal with farmers' many complaints about an overabundance of deer and the resulting damage to agriculture, and conversely, deer hunters' alarm at the shortage of deer. I was to be part of a four-person team. Bill Longhurst, who already had his doctorate from Cornell, would head up the statewide survey. Thane Riney, a naturalist trained at the Hastings Reservation, and Randall McCain, who was qualified in forestry and range management, would do an intensive study of the Jawbone deer herd, located on Jawbone Ridge north of Yosemite. As part of the deal we all had to move to the Sonora area in the Sierra foothills, with our families. I realize now that this was a trick on Starker's part to get us out of his hair. He feared, I believe, that if we stayed in Berkeley we would be in his office all the time and would interfere with his other work. In Sonora we could wait for periodic visits to get his advice.

Elizabeth and I, with our two-year-old daughter, Sandra, found a place in a small town above Sonora, Twain Harte, named in honor of Mark and Bret. There we occupied an uninsulated summer cabin during the worst winter in living memory. While I was away on the survey, heavy snows fell, the pipes froze, and Elizabeth had to get water by chopping a hole in the ice covering a nearby pond. Our daughter did not like snow all that much after she discovered that she sank in over her head. We finally had to migrate down-mountain to a place called Camp Sunshine, which we hoped it would be.

Bill Longhurst and I got along well. We worked together in the sagebrush country of the northeast. However, once we had our methods coordinated, we divided up the state and only occasionally came together on the same area. On the Jawbone side of the study, Thane and Randall got along about as well as the Israelis and Arabs. Thane did an excellent job of focusing on the deer, whereas Randall concentrated on the habitat evaluation side of the study. The results of this work were published, digested, and put into operation by the Department of Fish and Game. There is no need to repeat them here.

For me the survey was a rare opportunity to visit all the wild country in California. I watched antelope and deer on the sagebrush plains of Modoc County, and coyotes in the Warner Mountains. I ran survey lines in the dense ancient redwood forests of Del Norte County, looked for deer in the Sierra and coastal ranges, the national forests of southern California, and for deer and bighorn sheep in the Anza-Borrego desert. The survey made it possible, later on, for me to write *The Destruction of California* based on first-hand experience with most areas of the state.

We discovered that in the available deer range, most areas of deer habitat, there was an overabundance of deer, with a strong likelihood of a serious crash or die-off from malnutrition and disease unless their numbers were reduced, preferably by liberalizing hunting regulations. It did not take mathematical genius to show that with a polygamous species such as deer, you could not exercise population control by hunting only the big bucks with forked antlers. In other words, does had to become fair game. When Starker presented these findings to various audiences throughout the state, the trouble really began. It was my first experience with presenting scientific findings to people conditioned from their earliest years to believing the opposite.

From the reactions of the deer hunter associations, you would think that proposing a doe hunt was equivalent to human matricide and infanticide in one package. Does were sacred: if they were hunted, deer would surely become extinct. There was a widespread belief that deer were scarce, and many deer hunters stated emphatically that there were hardly any big bucks left anywhere. Each hunter believed himself to be an expert on wildlife and would not listen to any nonsense from a bunch of college boys.

In fact, some doe hunting was approved by the department and the Fish and Game Commission. It reduced deer numbers in some areas and did seem to stimulate reproduction and survival, so that some of the best buck hunting followed in years after the doe hunt. Nevertheless, deer hunters continued to believe the opposite and maintain to this day that the sacrilegious doe hunt brought an end to decent deer hunting. Regrettably, the organized hunters had their way. Through influence in the state legislature, doe hunting was prohibited without consent of the county supervisors. In the rural counties, such consent was as easily obtained as a furlough from hell, so liberalized hunting laws did not go into effect. The results were as we predicted. During the middle 1960s, deer populations crashed to a much lower level. But you get no praise for predicting catastrophe. Ask Cassandra.

During much of this turmoil I was on the sidelines, being a lowly graduate student. My brother Bill, however, was in the thick of it. At that time he was in charge of big game management for the Department of Fish and Game. His background in range management put him in the forefront in deer habitat evaluation. He had trained Longhurst and me in range condition survey techniques and was one of the top people in his field. Nevertheless, he incurred the wrath of deer hunters. More than thirty years later, when I was a member of the Fish and Game Commission, an old hunter from Modoc County called me down for having recommended doe hunting back in the 1950s, mistaking me for my older brother.

I would like to say that with the enlightenment gained from the decades that have passed, such controversies are a thing of the past. Unfortunately they are not. Now, however, wildlife scientists and managers tend to bunch together with deer hunters, duck hunters, and even fisher people, surrounded by emotional crowds of animal lovers who regard anyone who would hunt or fish for sport to be the equivalent of a cigarette-smoking, crack addicted psycho-killer. Sometimes it is necessary to have police guards when the two sides meet for a friendly exchange of points of view.

The basis for the problem seems to lie in one major difference in viewpoint. One side looks at the individual animal and considers its welfare. The other looks at the population and considers its probable future. In hunting, the individual is the target. It suffers and dies. Without hunting, or a heavier level of predation than occurs in most areas, the population increases to a level where many, most, or all individuals suffer and die. Yet the deer population is not a visible entity, since the black-tailed and mule deer are not herd animals. Some of the anti-hunting group say, "Bring back the original predators." Others, who have seen at least a video of a predator bringing down its prey, shudder and say no to that. Some advocate birth control measures. Inject the animals with antifertility chemicals. Believe it or not, this has also been tried. It does not work. If you were to remove all humanity from the scene and allow natural processes to operate, including the return of all the native predators, some sort of fluctuating balance would eventually be reached. Chances are a lot of folks would not like it. Besides, humans insist on remaining part of the scene.

In the course of events while we were living in the Sonora area, pregnancy overtook us once again. With the statewide survey essentially completed, it became necessary to move back to the Bay Area. Housing, however, was a problem, and we ended up in a housing project in San Pablo. This was what we could call the pits. Our daughter Marlene was not brought home from the hospital into ideal surroundings, but we did our best to welcome her. I scrambled for a source of income, and fortunately there were several job opportunities. The best seemed to be a game biologist position in the wilds of Siskiyou County, home of Big Foot. After much discussion we decided to go for it. But that night neither of us slept well. By morning we both said to hell with money, let's go for a Ph.D. instead. So we did.

The idea of my going for a Ph.D. had become increasingly imperative to both of us. In many respects I had seen Starker Leopold's life as being close to an ideal situation. He had done his doctoral thesis on wild turkeys and had been able, for example, to put out his book on the wildlife of Mexico and do the fieldwork on which it was based—as well as his later study on the California quail—while maintaining his position as a professor. I was by then familiar with the restrictions on personnel in state Fish and Game, Forest Service, and other agencies, as compared with Starker's ability to follow his own interests. I knew that a master's degree would be good enough for an agency job, but to go on with research and writing, a Ph.D. was essential.

Fortunately, Starker Leopold also thought it was a good idea for me to go for a Ph.D. and pointed out that Harold Biswell of the forestry school had a state Fish and Game Department project in Lake County that involved a deer study and could lead to a doctoral thesis. One of Starker's graduate students, Dick Taber, was already involved and it seemed there was funding enough for me also.

Achieving a master's degree or a doctorate meant that you had to decide on a thesis (dissertation) subject, carry out the necessary research and then write the results of your studies in an acceptable form. You had to make an original contribution to knowledge in your field (which could mean a lot or a little). Many a graduate student gets hung up on writing their thesis and remains in that limbo of "all but the thesis" forever. Fortunately, I had a major professor, Leopold, who wrote beautiful English and Spanish prose and was a great influence on me.

During my deer study years, all of my writing was on the technical side—papers published in refereed journals—for example, the *Journal of Wildlife Management.* It was not until my teaching began that I developed the confidence to write papers and books more general in scope, concerned with the human dilemma on a small planet. In this I also had my guides: Aldo Leopold, Carl Sauer, Frank Fraser Darling, and many others. But I will discuss this more later.

Switching work from the whole state of California to Lake County—only an hour and a half's drive north of Berkeley—meant that I could stay at home more or that the family could go with me into the field more often. And so we needed to find a house in Berkeley.

In 1950 we were able to buy a house in Berkeley on McGee Avenue. It was a small two-bedroom place, one story in front and two in back. The ground floor in back was convertible into two rooms and, over time, Elizabeth did the work of converting it and putting in a stairway to the upstairs kitchen. Unbelievable today, this place cost only $10,000, which we could cover with a low interest, California veterans' loan.

Working with Dick Taber on the Lake County project and operating out of our house in Berkeley, I now began an intensive study of a deer population. I stayed within a small area and became acquainted with the deer that lived there. Some of them I grew to know quite well. They had different personalities, ranging from calm and phlegmatic to hyperactive and excitable. We were able to do one thing that many other researchers had not: to trap and ear-tag deer for individual recognition at a distance. That brought some exciting moments in its early stages when we used large corral traps. One of us had to go, like a gladiator into the arena, tackle the deer, subdue it, and truss it up for weighing, measurement, and tagging. It was rough on the deer, but neither we nor the deer received any real injuries. Oddly enough, some deer became addicted to this exercise, "trap happy" we called them. They kept coming back for the alfalfa or apple bait and put up with the whole wrestling match several times. Fortunately, smaller traps were devised that required less strenuous activity. We were glad, however, when tranquilizing drugs for deer were developed.

We found that deer in the chaparral country of Lake County, unlike the migratory deer of the Sierra, spent the entire year in a relatively small area, often no more than a half-mile across. That area would contain dense brush or woodland for escape cover, more-open areas for feeding, and relatively cool, shady areas for avoiding the hot summer weather. Our principal aim in the study, however, was to examine the effects of brush fires, both wildfire and limited prescribed burns, on the deer population.

Wildfires occurred infrequently in this type of landscape but were usually devastating when years of fuel accumulation had built up. After a fire, many of the shrubs would regenerate quickly from their root crowns, whereas others would produce abundant seedlings. The burned areas were nutritional Edens for deer. The population would increase rapidly to more than ten times that maintained in old-growth brush. After five or six years the nutritional value of the brush would decline and much of it would have grown out of reach of the deer. A die-off would take place and deer numbers would fall again to the old-brush level.

The ideal habitat for maintaining a moderately high level of deer over the years was an open shrub land with brush patches and grassland intermixed. This could be maintained over time with periodic limited prescribed burning in the spring or fall, followed by reseeding to grass. However, what is best for deer is not necessarily best for other species. Some seem to prefer the old brush: wren-tits are an example. The best answer for all concerned may be a landscape mosaic, with large enough areas of old brush, young sprouting brush, and grassland. Dick Taber and I have gone over all this in great detail in "The Black-Tailed Deer of the Chaparral," a California Department of Fish and Game bulletin.

During our Lake County deer research, in 1950, I had the opportunity to meet a man who was to be a major influence on my deer research and on my life, though I had no idea at the time of the long-term effects of our meeting.

During most of his stay in the United States he was known as Frank Darling. In later years he insisted on Frank Fraser Darling, and in time he received a knighthood. Regardless of title, Frank was a major influence on my work. He was, in my opinion, the leading British ecologist of his time. The work that first had a great effect on me was *A Herd of Red Deer* (1937), an elegantly written account of the years he had spent studying red deer in the Scottish highlands. Without benefit of computers, dart guns, radio transmitters, helicopters, or any of the high tech gear of modern wildlife researchers, he discovered more about the ecology and behavior of red deer than anyone else had in his day. European red deer are clearly related to our elk and not similar to our mule deer or black-tailed deer.

Frank had come to Lake County to look at the deer study Dick Taber and I were conducting. He gave us encouragement to carry on, which we needed at that time. Frank was well over 6 feet tall and heavily built. He tended to speak slowly, and indeed, to take his time about most things. "Idleness is a virtue," he told me. "Not laziness, which is different, but being able to sit back and let your mind play with ideas and concepts." His thoughts on the virtue of idleness had more than a little appeal to me, but the opportunity to practice it seemed limited in the days to come.

In 1953 I had finished all but one of the required courses for a Ph.D. The remaining hurdle was a field course in marine biology that was offered in the summer at the Dillon Beach Marine Station on Tomales Bay in Marin County north of San Francisco. It was normally taught by Joel Hedgpeth, who authored the revision of *Between Pacific Tides* (1939) by Ed Ricketts and Jack Calvin, considered the outstanding guide to intertidal marine life in California and northward. But this summer Joel was on leave and his place was taken by a visiting professor from the eastern United States who had one problem. He was extremely shy, and the sight of a classroom full of students waiting for the word seemed to frighten him. Most of his lectures were given to the blackboard and audible only in the front of the room. In the field, if you could catch him, he was OK and obviously knew his material. Still, we all had Hedgpeth's book and the whole intertidal zone of Tomales Bay to learn from, so some of us learned a lot.

My longtime friend from our deer studies, Dick Genelly, was with me in this course and we both brought our families along to stay in a house in the redwood forest at Occidental, not far from the beach.

Nothing went well for our professor. He was beset by a series of incidents. On one day, he split the class up into small groups, each to run transects from the high tide line out to the deepest water they could work in at low tide, recording all the species encountered and their abundance, starting at Tomales Point, a sheer, rocky headland at the mouth of the bay, and moving from there back into the bay.

We all started in with enthusiasm at low tide in the morning and soon became very engrossed. The Tomales Point folks, however, forgot about the tide coming in and found themselves trapped when it did. Fortunately, one of the students was an experienced rock climber and had the necessary ropes and equipment available. With the help of other students, she was able to get everyone safely to the top.

The second event was when we went to learn about how to carry out surveys in deeper water. For this we had the marine station's research vessel, the *Bios Pacifica,* along with its skipper. We all set sail for the mouth of Tomales Bay and reached it in a short while. Then, however, we ran out of gas and with a final sputter found ourselves at the mercy of the tides, winds, and currents. Fortunately, there was a dinghy. Dick Genelly and I volunteered to row ashore and talk one of the farmers into selling us 5 gallons of fuel. This we did and managed to get the engine running and the *Bios Pacifica* back to its harbor.

Obviously this summer course hardly qualified me as a marine biologist. Nevertheless, I retained an interest, not so much in the open oceans as in the edges of the sea, the intertidal zone where lands and water meet and beyond to the depths in which corals thrive to build their reefs and create new islands in tropical seas. This interest, and my admittedly limited experience, led me to represent the International Union for the Conservation of Nature and Natural Resources (IUCN) many years later on projects concerned with oceans and islands.

In 1953 we had finished much of the fieldwork needed for the deer population and chaparral fire study we had been involved with, and funds for further support were running low. With my course work for the Ph.D. finished, I needed only to write my thesis. Though I was being supported by a fellowship from the National Science Foundation and was off the Fish and Game payroll at that time, I felt the need for a "real job." Unfortunately, the only teaching position open was with the biology department of the University of Minnesota. I am not sure why I felt the need to apply for it, but I did and was hired. By then, my daughter Marlene was three years old, while Sandra had reached the age of seven. We left my mother to look after our house in Berkeley and all took off in our ancient Buick to drive across the country to Duluth, Minnesota.

In crossing Wyoming we encountered a peak in the ten-year cycle of the snowshoe hare population. There were bunnies everywhere and it was difficult to avoid running over them.

In Minnesota I was asked to teach invertebrate zoology, comparative vertebrate anatomy, general zoology, dendrology, ecology, and conservation of natural resources. I also had to write my Ph.D. thesis. We arrived in the autumn, but winter soon set in. Snow was abundant and the air temperature fell to 34 degrees below zero Fahrenheit. Our car had no heater. I don't know how we survived, but we did. I became a quick learner on the plants and animals of the eastern hardwoods and pine forests. Then, to our joy, a job materialized at Humboldt State College teaching only wildlife management. I applied and was hired. Still in our old Buick, we made it back across the U.S. to a house provided for us in Arcata, on the campus of Humboldt State.

Perhaps because of the Minnesota weather, pregnancy overtook us once again. That would have been a problem in itself, but on top of it the two children came down with a sickness and passed it on to Elizabeth. It proved to be mumps, which Elizabeth had escaped as a child. It hit her hard. With advanced pregnancy and her face and neck swollen with mumps, Elizabeth was not really ready for a cross-country trip. However, when she talked to the doctor about her situation, he recommended that we get in the car and "drive like hell for California." We did that. In Wyoming, we stopped for a meal. Elizabeth had her lower face and neck covered with a scarf. Then Marlene piped up in a clear voice, "If they knew what you had, Mom, I bet they wouldn't serve you."

We expected Lauren to arrive on the Fourth of July, and we made that deadline. Lauren, however, had other plans and waited until August 3, 1954, to be born.

9

# Arcata

I started teaching at Humboldt State College in the fall of 1954 and continued to teach there until the summer of 1959. My work as an author had already begun. In 1952 I published "Methods for Estimating Deer Populations from Kill Data," which was from my master's thesis, and in the same year, with William Longhurst and Starker Leopold, the *Survey of California Deer Herds.* My Ph.D. thesis on "Ecology and Social Behavior of the Columbian Black-Tailed Deer in California" was published in the *Journal of Mammology* in 1956 as a coauthored paper with Dick Taber. And in 1958 Dick Taber and I saw publication of "The Black-Tailed Deer of the Chaparral." Now I began writing books.

As I have mentioned at the start of this account, I had no intention of becoming an author when I first started to pursue wildlife biology as a career. I was led into it gradually. First Vinson Brown, founder and owner of Naturegraph Publishers, wanted someone who could write and illustrate to do a book to be entitled *The Pacific Coastal Wildlife Region.* It would be part of a regional series of books that would serve as guides to the common plants and animals of the northern coast of California. He came to see Chuck Yocom, in charge of the wildlife program at Humboldt State College, who was a capable botanical illustrator. Chuck agreed to do the plant section of the book and suggested that Elizabeth could do the animal section. She agreed, and I tagged along to write a brief account of each species. When the book was published in 1957, she insisted that my name and Yocom's be listed as authors since I "had the degrees" and reputation. Her name appeared inside the cover as the illustrator, even though she had done most of the work.

Next I was approached by the local representative of John Wiley and Sons, New York textbook publishers, and asked if I could write a book on conservation of natural resources. Since I had taught such a course in Minnesota and was doing it again at Humboldt, I showed him my lecture notes and said I had essentially written it and needed only to get it into publishable form. After some negotiating back and forth and a lot of editing and some changes and additions, it was published in 1959 under the title *Environmental Conservation.* I had a lot of difficulty getting the title accepted: the term "environmental" was at best little known, objected the manuscript reviewers, and "conservation," to most people, simply meant keeping, storing, or even banking. However, thanks to my editor, Earl Shepherd, generally known as Shep, my suggested title remained.

I was fortunate to have my notes from Carl Sauer's class at Berkeley, since I relied on them to a great degree. I was also fortunate to receive a copy of *Man's Role in Changing the Face of the Earth* (1956), to which Sauer and many others had contributed. I actually read that huge tome from cover to cover and referred to it extensively. Once again Elizabeth contributed all the illustrations except for photographs but did not want her name put forward.

During my years at Humboldt State College I had frequent opportunities for research. Within easy reach of the Arcata campus of Humboldt State were a variety of different ecosystems. Humboldt Bay itself was a wintering ground for many species of waterfowl, and great numbers of many different species passed through the bay region in the spring and fall migrations. The bay was closed off from the ocean by the northern and southern sand spits, of which the northern was sufficiently wide to support extensive sand dunes and plant communities ranging from intertidal through to mature forests and freshwater marshes. Inland from the bay, redwood forests dominated, from old growth or ancient stands to recently clear-cut areas and some that had been logged and burned during pioneer times. Across the coastal mountains the vegetation shifted to Garry oak woodlands and savannas, where in early morning or late afternoon hundreds of deer could be observed. Not far away also was Prairie Creek state park with its herds of Roosevelt elk. Thus in many respects Humboldt State was ideal for classes in ecology or wildlife biology in which the students could have first-hand experience in the field with the concepts and ideas being discussed in the classroom.

I guess my interest in sand dunes relates back to my childhood in San Francisco at a time when most of what is now the

Sunset District was covered by sand dunes. My childhood companions and I were able to pretend we were the French Foreign Legion or, alternatively, Bedouin raiders waging battles across the Sahara Desert, though we actually were where the future grids of streets and avenues would be located and the dunes would disappear.

Starting in 1954, I became particularly interested in biotic succession on coastal dunes and with the help of my students established transects and plots in the various dune communities. We could observe how dunes were stabilized and also how moving dunes invaded and set back successional communities and even long-established beach pine or spruce-fir forest stands.

I was interested in vegetation structures, the classification of biotic communities, and the edges or ecotones between the communities. I continued with my interest in deer and began regular population counts, determining the sex and age structure of the abundant population in the oak woodlands and savannas. With the elk population at Prairie Creek, my interest focused on food habits and the extent to which the elk made use of native grass and forb (broad leaf herbacious plants) species as compared with the invasive annuals. All of this led to a great accumulation of data, which might have led me to write some useful papers had I not gone off to Africa, and later to Washington, D.C.

10

# Conservation by Slaughter

Since I grew up with Roosevelt's *African Game Trails* on hand, I may have been fated to go to Africa. But early in my wildlife career this had seemed impossible. Yet from the time the statewide deer survey ended, I had kept in touch with Thane Riney. He had moved to New Zealand, where he became involved with their deer problems. "Problems," because deer and other mammals, except bats, were not native to those islands. They were currently doing damage to native vegetation and more certainly were competing with sheep farming. Riney was a highly mobile person and—thanks to the Fulbright Act, which provided for American scholars to do research overseas—in the late 1950s had moved from New Zealand to the crown colony of Southern Rhodesia (now Zimbabwe). The National Museum of Southern Rhodesia had requested three Fulbright positions from the United States. Thane filled one of them and wrote to ask if I would be interested in coming over to occupy one of the other spaces.

It made no sense for me to leave Humboldt at that time. In my fifth year I had been promoted to associate professor. I had tenure, and I was in charge of the wildlife program. I earned almost enough to live on. I had two books published, and a lot of interesting research under way. Elizabeth was doing well as a freelance artist, with all the necessary connections. Our daughters were aged four, nine, and thirteen and were settled into their lives.

Elizabeth and I had long talks about the idea of moving to Africa. In this decision, as with all my other career moves, she backed me totally. I can't explain why she agreed to pick up our three children and go off to Africa with me, but I know she did not object or I would never have decided to go. Elizabeth was always adventurous and enormously resourceful. If she did not have a needed tool, she would make one. If there was no chair, she would build one. If there was no bathroom, she would dig one. After a childhood spent in the Australian bush, she did not find the African bush as big a stretch for her as it would have been for most people. And as for the children, Elizabeth was always there to teach them how to create, adapt, draw, sew, write, paint, read, hammer together, saw apart, build, invent, understand, or develop as the need arose. In short, she was an unusual woman, and we decided to go to Africa.

So I applied for and received the Fulbright grant, resigned from Humboldt in order to lay hands on my retirement fund, received a royalty advance from my textbook publishers, and began preparation for departure.

In retrospect I wonder at our boldness. I'll blame it on karma. It changed the whole direction of our lives and led me into the field of international conservation.

In 1959 air travel was neither as fast or as convenient as it is today. Our flight to London with three children was lengthy and uncomfortable. London in 1959 was still wrapped in imperial glory, not what it is today. We stayed in a quaint old hotel in Kensington called The Prince's Lodge. Which prince was not explained. One of the high spots of our visit was a formal reception for Fulbright scholars held by the chancellor of the University of London. Elizabeth and I tried out C. Northcote Parkinson's instructions on how to meet the "important people" at such an affair, which told us when to arrive, and how to move across the room to reach the appropriate place at just the right time. It worked. We arrived at square E-7 or whatever it was at the designated time, and sure enough there were the American ambassador, the chancellor, and all the other top brass.

We also spent some time at the London Zoo, where we saw some of the African species we would be working with, many of them for the first time. In London we met the third member of our Fulbright team, Archie Mossman, also a former graduate student at Berkeley but more recently teaching at the University of Wyoming, here with his wife and three children. Much of our time, however, was spent in Fulbright formalities telling us what Rhodesia would be like and how to get along with the people there.

The flight from London to Salisbury was even longer, or so it seemed, than our flight to London. Its low point was a stopover at Khartoum in Sudan. We saw only the airport, where the temperature was near 120 degrees and air-conditioning nonexistent. The fact that our four-year-old found a large cockroach in her orange juice added to the discomfort. Salisbury at that time was a beautiful city with tree-lined streets blooming with purple jacaranda and the scarlet of flame trees. But we stayed only overnight and proceeded to Bulawayo, another city of broad boulevards and a similar array of flowering trees.

We were met in Salisbury and again in Bulawayo by Reay Smithers, the director of the National Museum. In Bulawayo he provided us with his flat, not far from downtown in a four-story building known as Warnborough Mansions. The flat was mostly furnished and would help us settle in. Mossman opted for a rented house in the suburbs. But neither Archie nor I did get a chance to settle in: the next day we were next driven for many hours over a one-lane highway to Wankie National Park. The highway had two paved strips, each about a foot wide, for the car's wheels. The rest was dirt. On this we learned the game of Rhodesian roulette: two cars come head-on at each other, each holding to the paved strips. The one who moved off the pavement first lost the game.

In Wankie we met Thane Riney and his New Zealand fiancée, Anne. Their headquarters were in the main campground where they occupied the round, adobelike huts with thatched roofs known as rondavels. Here for the first time we heard lions roar and the trumpeting of elephants. We saw the game herds—giraffes, zebras, buffalo, elephants, wildebeest, warthogs—and the great bounding herds of impala. Not far away in Victoria Falls National Park we encountered our first hippos and crocodiles and saw the great rolling waters of the Zambezi as it moved inexorably toward the spectacular falls. Africa was not a disappointment. It was living up to its publicity.

However, as time went by the human aspect of our work

began to dim the brilliance of the natural scene. I do not recall what kind of studies Thane Riney had in mind for us to do, but I do know they had little appeal, if any, for either of us. We both decided we must set up an independent project, away from Wankie, one that would examine the potential of African wildlife to compete economically with the livestock industry.

We traveled considerably around Rhodesia and talked to many people in searching for an ideal study area. Finally, Reay Smithers directed us to a ranch belonging to Ian and Alan Henderson located in the low veld, 130 miles south of Bulawayo. The Hendersons had divided their ranch into two properties, Manyoli, being developed for cattle; and Doddieburn, being left for wildlife. Doddieburn, a 50-square-mile area, seemed ideal for our studies, and the Hendersons very generously agreed to allow us to carry out our work on their land.

Our plans for this study had been prompted by ideas advanced by Frank Fraser Darling, E. B. Worthington, and Bengt Lundstrom, who indicated that native wildlife would be more productive and have greater economic value than domestic animals. They were backed up by more recent studies by Bill Longhurst in Uganda, and Lee Talbot in Kenya, as well as by data from Rhodesia stemming from efforts to eliminate wildlife.

In Rhodesia wildlife was being killed off on a massive scale, both to control the spread of the tsetse-fly-borne livestock disease known as *n'gana* and also to "clear" land of those species deemed likely to compete with cattle for forage, along with any predators, from jackals on up. In addition to *n'gana* it was feared that wildlife would spread hoof-and-mouth disease and the malignant catarrh reputedly carried by wildebeest.

Our argument was that the government and private landowners were spending a lot of money trying to eliminate the more valuable resource, wildlife, in favor of the less valuable, cattle. We believed from the start that the multiple species of wildlife would feed on a much broader spectrum of vegetation, from giraffe at the treetops to warthogs at the grassroots, than domestic cattle, and could therefore sustain a higher biomass (weight) and productivity from the same area.

Henderson's was probably the only place in Rhodesia where our study could be conducted. Much of the original wildlife was still present. A few species had disappeared long ago—sable, roan, rhinos, hartebeest. Other species had been reduced in number, particularly the larger predators. Still, there were enough lions and leopards around to liven up our fieldwork. The array of hoofed animals was a bit dazzling to an American used to working with much lower diversity. We hoped to be able to take a sustainable crop of wild species and market their meat, hides, and other items of commercial value. Then we would run comparisons between the total weight of the wildlife crop, compared to the sustainable annual production of cattle, and also run the same comparison with the cash return from wildlife versus cattle.

First of all we had to determine the existing population of each species and evaluate the annual production of young (fawn or calf crop) and their rate of survival to adult size. This we did through various "census" methods and herd composition counts that indicated the ratio of males to females to young animals. Our methods were necessarily crude but conservative and have been described in the book *African Game Ranching,* and in various scientific papers that resulted from our study.

The one thing we did not want to do was overestimate the population or the sustainable yield and then see the numbers of any species decline after the first year of wildlife removal.

The area where we were to work was characterized by dry tropical deciduous forest. The climate was typical of the two-season tropics, a long, dry cold season in winter followed by a shorter wet season when insects became superabundant and travel was made difficult by muddy roads and trails. In the wet season the vegetation was luxuriant. In the dry, however, most trees shed their leaves and the landscape appeared arid. Most of the country was relatively flat, broken by stream channels, mostly running only in the wet season, and by kopjes, high granitic domes surrounded by great, wind-carved boulders. From the kopje tops we could see over the tree canopy and at least discern giraffes at a distance. Kopjes were home to small antelopes, called klipspringers. In some areas they supported large colonies of dassies, or rock hyraxes, who greeted visitors with permanently fixed smiles. Usually they were home to a troop of baboons. These water catchment areas held water after most streams were dry and therefore could support larger, evergreen trees such as figs, which were absent from the surrounding plains.

Though there were open grassy areas and some open brushy areas, much of the land was covered by a more or less closed canopy of woodland into which vision was restricted, particularly in the wet season. Consequently you could often be surrounded by animals you could not see unless they crossed the road. The only areas that resembled rain forest, with tall, vine-bearing trees, were the riparian forests along the major rivers.

The presence of lions and leopards affected my activities and attitudes. Though I felt compelled to run vegetation lines through dense brush and woodlands to determine species composition and use by wildlife, I became sensitized to threatening sounds. The sudden snort of a kudu could cause me to grip my rifle and prepare for the worst, until I remembered that it was not a large carnivore. In fact I never saw a lion or leopard at Doddieburn, though a lion's roars enlivened evenings by the campfire. Mossman and I worked separately on many occasions, so that I was often camped alone by my favorite kopje, entertained by a troop of baboons noisily settling in for the night in a large fig tree. They watched me with as much interest as I watched them, but only I had binoculars.

We had no trouble going on with our censuses, composition counts, and other data gathering. When it came time for killing and marketing animals, however, we had to get permission from the Game Department, whose chief, Archie Fraser, was not inclined to look kindly on a couple of Yanks who were asking to break all the rules in the wildlife protection book. We were saved from this problem by the arrival of Peter Crowcroft, then curator of mammals at the British Museum of Natural History. I had become associated with Peter during our stay in London and had told him of our plans. He was interested in our efforts and also in need of new mammal specimens for the museum. He decided to work with us. He carried enough British scientific authority, as compared with our American brand, to sway Archie Fraser and obtain permission to proceed with game ranching.

To keep us honest, a young Rhodesian game ranger, Alan Savory, was assigned to work with us. Later a second man, Peter Johnstone, was assigned to our team. Both of these men were to gain greater fame after we left: Johnstone by operating a successful game ranch near Wankie; Savory by entering Rhodesian politics and later moving to the United States, where he started teaching holistic range management. However, both learned their first lessons in wildlife range management from what I had learned earlier from Arthur Sampson, Harold Biswell, Harold Heady, and most of all, my brother Bill.

During Crowcroft's visit, Mossman and I went with him to have a look at "Operation Noah," which was proceeding on the Zambezi River in the north. This was an expensive wildlife rescue operation brought on through the misguided damming of the Zambezi at Kariba. The Kariba dam was to provide power to the industries of Southern and Northern Rhodesia but flooded the homes of the native people who lived in the Zambezi valley as well as the natural vegetation and wildlife. To save wild animals from being drowned, game rangers traveled around the lake, immobilizing animals and transporting them to a new national park on the southern shore of Lake Kariba. But after we had traveled around the lake by boat, watched the game rescue operation, and witnessed the drowning forests, we concluded that the game rescuers were admirable but the whole endeavor was unfortunate. The trip was partly ruined for me when I was bitten by some unidentified tick and came down with some sort of tick fever that kept me under medical care for a couple of weeks.

When we had gone far enough with our wildlife population evaluation to feel safe in beginning cropping and marketing, the heavy work started. It involved hunting, shooting, hauling to base, skinning, cleaning, loading on trucks, and transporting the game to cold storage facilities at a Bulawayo market. Game meat sold for relatively high prices. Hides had good commercial value, and the price received per animal was adequate. However, everything from shooting to delivery to market had to be done quickly, since we had no refrigeration facilities at the ranch. In these logistic areas, cattle ranching had all the advantage, since all they had to do was round up the cattle, "put 'em in the cars," and say good-bye to them. In this end of the business I became less and less useful. To begin with, I didn't like killing animals and over time developed a flinch that caused me to miss most of my shots. Fortunately, Alan and Archie remained enthusiastic hunters, but we could not have gone far without the African game scouts assigned to work with Savory. They did most of the heavy and dirty work.

Originally we wanted to work in some relatively remote area of the country, with the Africans who lived there developing wildlife ranching as a means for supplying their needs and bringing in some cash income. However, we were told early on that this was out. Perhaps those in charge did not want Americans stirring up the local people with the idea that they could be in charge of wildlife on their land. Such a scheme would eventually be put into effect, but only after a quarter century had passed and Southern Rhodesia had been replaced by the independent nation of Zimbabwe. By that time I was on the board of the World Wildlife Fund, where we agreed to provide several million dollars to help Zimbabwe get on with this project: I gained some satisfaction from telling Prince

Philip and Russell Train that in 1960 we could have done it for one-tenth the price. Back in 1960, however, we were told to work on European-owned lands.

Officially Southern Rhodesia was quite different from South Africa. Apartheid, in theory, did not exist north of the South African border, but in fact discrimination was deeply rooted. While we were in Africa, the civil rights movement was getting under way in North America, and Elizabeth and I would be challenged about it in almost every conversation that developed with Rhodesians. The two of us had problems with the idea that we were expected to have African servants to cook and clean for us. We both preferred to do the work ourselves. When we were in the field we also had seen no need for servants to put up tents, build fires, cook, and do all the things we were used to doing. In time, however, we were made to feel bad about depriving Africans of employment and were prevailed upon to hire Reay Smithers's former housekeepers.

The whole issue of racial discrimination came home to me when a man from the Northern Rhodesia Game Department was assigned to work with us at Doddieburn to learn our methods. Northern Rhodesia at that time was a British protectorate being prepared for independence (as Zambia) and, being further from South Africa, was less exposed to apartheid practices.

Sidney Simpelwe was from the Bemba people and was both well educated and of high social status in his homeland. In Southern Rhodesia, however, he was just another "boy." As long as we stayed at Doddieburn we could work well together, since the Hendersons and Alan Savory recognized his status. Elsewhere we had trouble. In Bulawayo he could stay in our apartment, but when he went out, he encountered discriminatory practices. In traveling to Wankie or Victoria Falls, we would not stop at hotels for a meal since they would not serve him in the dining room. We carried our lunches, prepared at home. Finally we gave up, and Sidney set up his base at Livingstone in Northern Rhodesia, just across the Zambezi River bridge from Victoria Falls. From there we could work in the Livingstone Game Park and Victoria Falls National Park, to practice various techniques.

When we had finished our first year of game ranching, we were able to compare our yield in biomass and cash with what the Manyoli cattle ranch should have produced. We found what we had expected. Wildlife definitely won in the competition. They could yield a greater return at lesser expense than cattle. Knowing as we did that scientific studies usually have little effect on the "real world," since other scientists are the only ones who see and read them, we proceeded to publish and publicize our results as widely as possible.

There was no doubt that the Hendersons were convinced, particularly since there appeared to be just as much, or more, wildlife present after the cropping as there was before, and there was no change in the animals' wariness or visibility. Despite opposition from the organized cattle ranchers, the Hendersons went ahead with game ranching. Over time they discovered there was more money to be made at less cost from organized safari hunting in place of the cropping and hauling that we had done. With this in mind, they began a program of wildlife restoration, bringing back species that had disappeared in earlier times. The sable and roan antelope are examples. Before the civil war broke out, there were over two hundred game ranches registered with the Game Department, but I do not know how these have fared.

Our time in Rhodesia was not all devoted to game ranching. Elizabeth and I were both particularly interested in the Matopo hills, not far from Bulawayo. These were sacred sites for the Venda people and for the white Rhodesians as well, since the grave of Cecil Rhodes had been placed there at his request. There was also a monument to the British battle with the Matabele during the early days of colonization. The Matopos were like giant-sized kopjes, great granite domes into which water and wind had eroded caves. The caves contained paintings done by the original inhabitants, the Bushmen, who with simple strokes captured the form, spirit, and motion of the wildlife: giraffe, wildebeest, kudu, zebra, white rhino, lion, and so forth. The paintings were obviously done by artists who were totally familiar with and empathic to the animals they painted. Elizabeth carried her paints and canvas to many of these caves and was able to reproduce a number of works in their original colors. Regrettably, the Bushmen had disappeared, driven out by Bantu invaders well before Europeans entered the scene. Those who survived, in the Kalahari Desert, know nothing of their artistic ancestors.

We were also able to visit the ruins of Great Zimbabwe, where remarkable stonework remained in walls and a tower built by an early African civilization, well before the arrival of Europeans. It provided the name and sacred symbol for the new nation of Zimbabwe.

We spent some time at Wankie, where we were able to get acquainted with, and sometimes scared by, elephants. The greatest show was at Nyamdhlovu Pan, where large herds of elephants, giraffe, and buffalo would come to drink. However, my most unnerving experience came at the Tuli Circle on the border with Botswana. This was a dry-season concentration area for species from Botswana and from the Rhodesian low veld, attracted by the water remaining in the Tuli River. Here, for my first experience camping out in Africa, along with several others, I decided to sleep out on the ground as the Rhodesians were doing. I awoke in the morning to find fresh elephant footprints no more than 10 feet from my sleeping bag. While we all slept, the elephants had come and checked us out, wandering through our camp. I suppose I could take it as a sign of approval of our mission that they avoided stepping on me. I have felt some kinship with them ever since.

I should point out that all five of our family members wanted to be in the field, watching the wild animals or just wandering around. The three children were fascinated by what they saw in the national parks and at Doddieburn and will never forget Africa.

During what were to be our final days in Africa, riots broke out in the African townships surrounding Bulawayo. The violence shocked us, since we had not sensed any bad feeling among the Africans we knew, but obviously we were out of touch. I was camping by my favorite kopje when I heard the news over the radio that riots and burning were taking place and the army had been called out. My family was at home in Bulawayo. Archie and Alan took the view that it was not serious, but I loaded up my car and headed for home. I was apprehensive as I neared the city that I might be stopped and attacked, but there was no trouble. I reached home to find the place surrounded by armored vehicles and obviously serving as a base for the Royal Rhodesian Regiment. Upstairs all was well, though my youngest daughter, Lauren, was wanting to go out and join the riot. She could see people in the distance dancing around a burning petrol station and thought they were having fun. We all watched from the windows as the Queen's African Rifles moved into the township to restore order.

The riots changed our attitude. Previously we had traveled freely around the countryside, feeling at ease among the Africans. Now images of Kenya's Mau Mau and the bloody stories of Belgian refugees who fled into Southern Rhodesia from the Congo came to mind. To accelerate our decision to go home, Starker Leopold wrote to offer me a teaching job at Berkeley. We sold out, packed up, and headed homeward.

On the way home we made our first stop in Rome, where I talked about our game ranching experiences with the people concerned with wildlife in the United Nations' Food and Agriculture Organization (FAO). From there we went on to Paris, where I discussed our experiences with those involved with wildlife conservation. From Paris we went to London, where Peter Crowcroft helped us rent a flat in Chelsea from which we could take time to visit British conservationists and get some writing done. It was then necessary to get back to Berkeley and prepare to teach Starker Leopold's courses.

11

# Return to the United States

After our African experience we did return home, and I put in a year working with the University of California at Berkeley. I was Starker Leopold's replacement while he was on sabbatical leave.

We came home to a different scene. While we were overseas my brothers, Bob and Bill, had decided that my mother could not stay in our Berkeley house by herself or even with her sisters, Aunt Peg and Aunt Nell, and had moved her to a nursing home in San Rafael. Considering that all three were in poor health, and that Elizabeth could not take on the care of my mother, we had to agree with my brothers.

When I finished teaching in Berkeley and went back to Humboldt State, we rented a house in Eureka. This enabled Elizabeth to get on with her artwork with the Redwood Empire Art League and made me less accessible for all the minor problems of academia than I would have been had we lived in Arcata.

I continued teaching courses I had taught before, including a beginner course on environmental conservation and one on the biology and management of large mammals (big game). Archie Mossman had stayed on in Southern Rhodesia when we left, but Humboldt State was able to hire him after he returned to the U.S.A. Dick Genelly took part in the African experience by taking a Fulbright grant to go to the University of Salisbury. He studied mole rats, just to be different and noncontroversial. So Humboldt was becoming a center for African research.

In due time I was promoted to full professor and took over as chairman of the Division of Natural Resources. We were all settled into our Eureka house and had sent the oldest daughter off to university at Berkeley, not knowing she was moving into the midst of the student revolt of the sixties.

I will not write much about Berkeley in the 1960s since I must leave that to Sandra, who was there. There was not much in the way of political activity or student protests at Humboldt State in the sixties and we did not go to Berkeley often. Elizabeth and I did manage to get caught in a small riot and took in some whiffs of tear gas, but that was just a case of being in the wrong place at the wrong time and not a political statement.

My writing continued at this time and I published several papers, including "The Economic Value of Rhodesian Game" in 1960 with coauthor Archie Mossman; in 1961 I coauthored "Commercial Utilization of Game Mammals on a Rhodesian Ranch," and in 1962 I wrote "Conservation by Slaughter."

I also published several more books in the early 1960s. One of those who read my first textbook and liked the way it was written was Frank Fraser Darling, who had contributed also to the book *Man's Role in Changing the Face of the Earth.* When a representative from Macmillan publishers came to ask him to write a book on the state of the earth with an emphasis on wild country, Frank referred him to me. I agreed to give it a try, get away from a textbook approach and write for the general public. *The Last Horizon* was published in 1963. It received glowing reviews from the *New York Times* to the *San Francisco Chronicle* and was to be put forward in the *New Yorker* as a contender for the National Book Award. One reviewer in the sports section of a Texas newspaper said he didn't like it at all and wondered why he had to read it. Perhaps he spoke for the general public. The U.S. Agency for International Development (AID) picked up the book and sent it out in the form of a low-cost paperback to developing countries. I don't know if anyone in India or Africa read it. It certainly did not sell at all well. I attribute this, in part, to the fact that the enthusiastic reviews did not appear until after the bookstores had returned their surplus books to the publishers.

Meanwhile, John Wiley and Sons, through Earl Shepherd, asked me to write a book on wildlife biology. I agreed and drew heavily upon both Starker Leopold's teaching and his father's writing; Aldo Leopold's *Game Management* was my principal reference. *Wildlife Biology* (1964) received wide acceptance among wildlife departments in universities and also community colleges. By 1999 it had appeared in Japanese and Chinese. I was told that people in Japan picked it up not out of any great interest in wildlife but in order to teach English to college students. Apparently my simple English style appealed to them, and perhaps some ecology rubbed off as the students learned English.

From 1964 on I seemed to be accepted as an author. I was asked to write a small paperback on game ranching as part of a British series intended to encourage children to consider going into scientific fields of work. *African Game Ranching* was published by Pergamon in Oxford in 1964. Russell Train happened to read it and later told me it had influenced him to ask me to join the Conservation Foundation in 1965.

My return to Humboldt State College in the early 1960s also gave me the chance to carry out more of the field research that had interested me for several years, including my work with deer and the local sand dunes. This work was interspersed with various expeditions out of the Humboldt area.

When I finished writing *The Destruction of California* at Humboldt in 1965, I thought about doing a book on California's islands. I had visited Santa Catalina on vacation trips with the family and had been impressed by the natural beauty of the island and the sea around it as well as the amount of natural vegetation that had survived the onslaught of cattle, sheep, pigs, and goats during pioneer days and was still present. And there was more to Catalina than the resort town of Avalon, even including its impressive surrounds. Beyond Catalina were the other Channel Islands. I wanted to learn much more about them despite the fact that visitation was discouraged by the defense agencies, who controlled San Clemente and San Nicolas, the National Park Service, who controlled San Miguel, Santa Barbara, and Anacapa, and the private owners of the larger islands—Santa Rosa, Santa Cruz, and Santa Catalina. There seemed to be little written about the natural history and even the human history of the islands.

My serious field studies began in 1968, when the National Park Service offered to take me on their boat to Santa Barbara, Anacapa, and San Miguel, the three islands comprising the Channel Islands National Monument. On Santa Barbara, I was able to record the bird species present and make my first acquaintance with elephant seals. I found the female seals sound asleep on one of the island's few pieces of sandy beach. Being extremely careful not to disturb them, I slowly stalked up to them, taking photographs each time I stopped to avoid scaring them. However, I discovered that nothing seemed to scare them and I almost had to kick them to get them to move. Perhaps this somnolence indicates why they were nearly driven to extinction during the pioneer days of uncontrolled whaling and sealing.

Santa Barbara appeared to be still recovering from heavy grazing pressure from sheep and from more recently introduced rabbits. I did encounter the unique giant *Coreopsis* plants that had survived.

Anacapa Island acquainted me with the cholla cactus, which seems to leap at you if you go near it. It is densely covered with spines that are just waiting to hook onto your hide.

On Anacapa, I was shown the nesting California brown pelicans, at that time considered to be an endangered species because of the effects of DDT and its derivatives. The Channel Islands and the Farallons to the north are major nesting and roosting areas for a great variety of seabirds, all of them to some degree affected by the introduction of DDT into the marine food chains.

On San Miguel Island, I was turned loose to hike across the island and join the boat on the other side. I observed the great numbers of seals on an extensive beach near where I was put ashore. These were probably northern fur seals and California sea lions. Like the elephant seals, the sea lions had come back from near extinction during the nineteenth century. I had my first encounter with the San Miguel Island fox, a sort of miniature variety of the mainland gray fox, which is considered by some authorities to be a different species from the mainland populations. A different race, or genetically different population, occurs on each of the larger islands, excluding Santa Barbara and Anacapa Islands. The ones I encountered were not much disturbed by my presence, so I was able to get quite close to them. They were beautiful little creatures and I could understand why the Chumash Indians of the islands may have kept them as pets. Regrettably, their numbers have declined and they are now considered to be an endangered species. Though I was to make a few more trips to the islands and became acquainted with the Santa Barbara Natural History Museum and their island studies as well as some of the work done by the Smithsonian, my shift from California to Washington, D.C., and the Conservation Foundation led to my abandoning the plan for an island book. Happily, however, Allan Schoenherr, Robert Feldmuth, and Michael Emerson were able to put together an excellent book, *Natural History of the Islands of California,* much better than I could have done at any time. This was published in 1999 by the University of California Press.

In 1963 Elizabeth and I had an opportunity to spend the summer in Costa Rica. The excuse was that I could take a summer course there in tropical ecology and compare neotropical

1. Ma always wanted a daughter. Instead she had three sons; I was number three. California, 1921.

2. Pa, Bill, and Ma in Yosemite, ca. 1912.

3. With ranch dog at Aloha Ranch, 1926.

4. At Kaar Ranch with my cousin Jack Gobar, 1937.

5. Cowboy days at Kaar Ranch, 1937.

6. Lake Tahoe, 1937.

7. Australia, 1942. *(From left to right)* John King, Ozzie St. George, the author.

8. Elizabeth, 1943.

9. Wedding day, Sydney, May 1944.

10. Elizabeth, Sandra, the author, and Ma on Liberty Street, San Francisco, 1946.

11. Lauren, the author, and Elizabeth on a beach in Humboldt County, 1961.

12. Lauren, Marlene, and Sandra in Southern Rhodesia, 1960.

13. Giraffes at Nyamdhlovu Pan, Southern Rhodesia.

14. Elizabeth copying Bushman art, Nswatugi cave, Matopo hills, Southern Rhodesia, 1960.

15. The author at UNESCO conference in San Francisco, 1969.

16. Prince Bernhard and Queen Juliana of the Netherlands conferring the Order of the Golden Ark, 1978. (©A.N.P. Foto, Amsterdam)

17. My brothers, Bill (*left*) and Bob, ca. 1986.

18. Cabin in the woods, San Juan Ridge, Nevada County, 1980.

19. The author with David Brower, Lake Baikal, Siberia, 1990.

savannas, which seemed to support few species of large herbivores, with the African savanna, with its high diversity of large ungulates. We managed to hire a housekeeper to look after Marlene, thirteen, and Lauren, nine, who would stay at home in Eureka while Sandra, then seventeen, would be spending most of the summer in a live-in summer school experience with a Mexican family in Saltillo.

We caught a plane to San José, the Costa Rican capital, and there managed to rent an apartment in the center of town. The course was of great interest, since its participants were able to visit most of the major vegetation types in Costa Rica, from the tropical rain forest and coral reefs on the Caribbean coast to the high *cordillera central* with its strange *páramo* vegetation and the dry savanna region of the Pacific coast. As a class we visited by bus the volcano Irazú, seeing the coffee plantations on the lower slopes and the ash-covered areas near the crater.

There are two major crater rims on Irazú. The inner ring surrounds the active crater, bubbling and boiling away. The outer must date from a major eruption in the past. Near the outside of the inner crater stands the ruin of a building I was told was a former café and crater-viewing site, partially destroyed by some past eruption. During the class visit, the volcano was very quiet and we were able to go to the edge and look down.

I thought that Elizabeth would like to visit the crater, so we rented a VW on the weekend following the class visit and drove up the mountain. We parked in a parking area outside the outer crater rim. Nobody was in sight, and there were no other cars. We went over the outer rim and down inside to the ruined building. Maybe somewhere, some time, I had offended the goddess Pele, who is in charge of volcanoes. When we looked into the lower crater, the crater floor started up toward us. We moved back from the rim but then, seeing the smoke and ash rising up from the crater, decided we had better hurry. Neither of us had trained for running uphill, through ash, at an 11,000-foot elevation, but we had to try. In no time a huge black cloud of smoke and ash rose above us. Glancing up, I could see lightning flashes and pieces of molten lava flying through the cloud. Our chances of survival looked slim. But we kept chugging uphill and finally reached the outer rim, where a ranger greeted us and asked if there were others inside. There were not. The tower of smoke and ash climbed thousands of feet, and soon ash was to fall on San José. We had ushered in a major eruption and only wanted to be down the mountain and away from there. We succeeded. I think it was the most frightening experience of my life, rather like an escape from hell. Twenty-four years later I went to the top of the volcano Poas with a World Wildlife Fund group, but I did *not* look down into the crater.

Our Costa Rican visit came to an end when a phone call from my Aunt Peg informed us that my mother had died. We dropped everything and went home. I did not do the comparative study I had in mind. Fortunately, others have since done a more complete study than I could have at the time.

In the summer of 1964 we went back to Australia. It was Elizabeth's first chance to see her relatives since coming to America in 1945 and my opportunity to meet most of her siblings. Her parents and grandmother had died during the intervening years. This trip was made possible by Peter Crowcroft.

Since I had seen him in Southern Rhodesia and in London, he had moved to take over the Natural History Museum in Adelaide, in his home country. He had arranged for me to give several lectures at the University of South Australia and to meet people who were concerned with conservation in Australia. He invited us to stay with him and his wife, Gillian. Gillian was also a biologist, and had a long-term study of marsupial mice underway at the museum.

We obviously could not resist this opportunity and so left the summer behind to go into an Australian winter via Qantas Airline. I gave my lectures, including one on game ranching and another about environmental conditions in America and what was being done about them. We had the opportunity to see the mallee woodlands, which are unique to Australia and home to a variety of colorful parrots. We also visited the Flinders Range and at least the edge of the gibber desert.

We had the opportunity to be taken around the Flinders and the desert by a sheep rancher who had extensive land holdings or grazing privileges. It turned out he was intending to demonstrate his marksmanship to us by pursuing the big red kangaroos and emus in his Land Rover. When he caught up with one of these, he would shoot it with his pistol and kill it. He obviously had no tender feelings toward these animals, who were trying desperately to get away from us. He proudly showed us a pile of dead wedge-tailed eagles he had killed. His excuse was that the kangaroos and emus ate the grasses that would otherwise be available for his sheep and that the eagles killed his lambs.

This was obviously a place where game ranching could be tried, and in some areas the marketing of kangaroo meat and hides was already a well-established practice. Refrigerator trucks accompanied the shooters to the field. Equally clearly, game ranching could also be a cover for the extermination of any wildlife that competed in any way with sheep or cattle ranching.

When we went to Seaforth, on the beach in tropical Queensland, a large number of Elizabeth's sisters, brothers, in-laws, nieces, and nephews greeted us. They found it quite a joke to serve us a dinner of brush turkey, an endangered species of mound-building birds, which they had poached from a national park. I could understand another reason why she left home. But there was a beautiful beach and we were able to do a four-mile walk along it one morning, seeing neither people nor houses but instead whole battalions of soldier crabs.

In 1964 I was invited by Frank Fraser Darling to present a paper at a conference to be held the following year at Airlie House in Virginia entitled "The Future Environments of North America." This conference was sponsored by the Conservation Foundation and organized by Frank. Of course, I agreed to participate and flew off to Washington in April 1965. My paper was entitled "Man in North America" (it was assumed in those days that women were included in the category of "Man"). At Airlie House I had the opportunity to meet some of the top people in the environmental field, particularly Russell Train, the new president and chief executive officer of the Conservation Foundation.

I returned to Humboldt and was settled in again for the fall semester when I had a phone call from Russ Train asking if I would be willing to come to Washington and join the Conservation Foundation as a senior associate. I would be working with Frank Fraser Darling, Bill Vogt, Ed Graham, and others whom I considered the most interesting people in conservation ecology. At that time the Conservation Foundation was probably the most prestigious organization in the field of conservation, a sort of think tank with considerable influence at the governmental level. Being asked to join them was, for me at that time, almost like being asked if I would accept a Nobel prize.

Once again came the decision to uproot the family, leave a comfortable position, forget once more the field studies I had under way, and go off somewhere else. After driving across the southern United States, we arrived in the District of Columbia in one of those snowstorms that shut down the government and everything else.

We rented a house in Georgetown and managed to get Marlene and Lauren established in schools. Sandra stayed on in Berkeley to continue with her anthropology major at the university. However, we were reasonably confident that we would be staying on in Washington, so we sold our Berkeley house and bought a house in the district's Chevy Chase area. Once again we settled in. Our real estate agent for the house purchase became a friend and was quite impressed by Elizabeth's paintings, so she arranged a show and reception for her at their offices in downtown Washington.

12

# Influences and Efforts

Our move to Washington, D.C., in January 1966 brought me into closer contact with my longtime colleague Frank Fraser Darling, a friend and mentor who influenced me repeatedly throughout my career. Frank's work in Africa had inspired me to get involved with African game ranching. He had pointed out that the use of wildlife in place of cattle would bring better economic returns to the African people over the long run, without the heavy habitat damage that accompanied livestock grazing. He later chided me about the emphasis in our game ranching studies, saying that he had not recommended commercialization of wildlife, but rather its direct use by the resident African population. So did we, but we were not allowed to carry out such a study.

I had the good fortune to join the Conservation Foundation while Frank was still vice president in charge of research. His health was beginning to be impaired by then, but his mind was certainly active. I sometimes traveled to his place, Shefford Woodland House, near Newbury in England, and later to his home territory in Scotland where he lived for a time at Drumnadrochit, near Loch Ness, and to his house near Inverness. The latter was not far from the New Age mecca of the 1970s, Findhorn. Frank had never heard of it. We put together a book called *The Environmental Revolution* in 1967, before Max Nicholson published his book with the same name. Ours was prepared as thirteen episodes for National Educational Television and was based in part on interviews with America's leading environmental philosophers, including Lewis Mumford and Carl Sauer. However, as it was being written in final form, NET changed its chief executive officer and the new chap didn't like it. "Too historical, not suited to the public interest of today," he decided. Since they had paid for it and owned it, it was never published or aired.

Frank was well able to combine sociology and economics with ecology, as he did in his *West Highland Survey* (1955), and later with J. Morton Boyd in *The Highlands and the Islands* (1969). His final work was the Reith Lectures of 1969, delivered over the BBC and published in 1970 as *Wilderness and Plenty.* In one lecture he stated that "the biggest problems facing the world today are the continuing rise in human population, the continuing rise and diversity of pollution, and finally, the increasing difficulty of preserving examples of the world's natural ecosystems with their species of plants and animals."

These problems that Frank Fraser Darling called to our attention in 1969 remain the biggest problems today, thirty years later. These are inescapable and must receive far more attention than they have thus far.

I remember his visits to our house in Washington. After due

consideration, he said, "If I didn't have money, I'd probably live as you do." We decided that was a compliment. However, he did have money, and his house was laden with treasures, from fine Oriental rugs to the best English china. I also recall when, after he had given a talk on the need to limit population growth, somebody asked him, "How can you talk about limiting populations when you have four children?"

"It's all right," he said. "I've had three wives."

Frank approved of some of my books, but he was always at me to do better. "You have it in you," he said. I explained about publishers and deadlines. "That's not the way you do it," he said. "Write the book, and when you feel it's as good as you can make it, then take it to a publisher."

So I explained about living from one royalty advance to the next, and the shortage of money. He had not thought of that.

My work as an author continued at the Conservation Foundation and, as I was asked to give many lectures to various audiences, I wrote many short papers. Sometime earlier Peter Ritner, my editor at Macmillan, had asked me to write a book similar to *The Last Horizon* but focused on California. The result was *The Destruction of California* (1965). I argued against the title, which I felt was too negative, but I lost. This was to be the most publicized and, in paperback, the biggest seller of my books. It put me into the category of California historian, which I am not, even though I continued to write shorter papers in this area. Once on a roll an individual tends to keep rolling, so I somehow continued to produce books and Elizabeth continued as my principal editor, contributor, inspirer, and illustrator. She became interested in photography and for our final book with Macmillan, concerning Florida, *No Fur-*

*ther Retreat,* she had one of her photographs on the dust jacket and chose other photos and drawings to go in the text. She finally let her name be used as the illustrator on the title page, but she was with me for most of the fieldwork, including launching me into scuba diving, in which she did not wish to participate.

*The Destruction of California* came out shortly before we left Humboldt and moved to Washington, D.C. The confluence led some to say I was driven out by the timber industry, which dominated the economy of the northern coast at that time. While it is true that I was more than a little critical of the logger/lumberjack mentality and the devastation the industry was bringing to the ancient redwood forest, I was never threatened by timber industry people. I doubt that most of the locals had even seen my book, let alone read it. At that time there was not a single bookstore in Arcata or Eureka and certainly none in the smaller towns. In later years I did encounter many people who asked me to write a follow-up to *The Destruction.* And so I finally came to write *California's Changing Environment,* published by Boyd and Fraser in San Francisco in 1981. It was part of a paperback series promoted by California historians.

*Wildlife Biology* went on to a second edition in 1981, whereas my old *Environmental Conservation* went through many reincarnations until its final and fifth edition in 1984. I was subsequently asked to do a third edition of *Wildlife Biology* and a sixth of *Environmental Conservation,* but I had to decline. I was tired of writing the same book, even though I had new information and many changes had occurred since 1984.

The conference in 1968 held by the Conservation Foundation on ecology and development led to the need for still another book, to be entitled *Ecological Principles for Economic Development.* I was with IUCN at the time it was being written. My coauthors were John Milton and Peter Freeman, who were both with the foundation. It was published by the British branch of John Wiley in 1973 and subsequently translated into other languages.

13

# Too Many, Too Much

One of the people I worked with in Washington was William Vogt. I had known of his work on human population since 1948, when his book *Road to Survival* was published. It was a startling book, and one that forced the media to pay attention and respond. *Time* magazine devoted most of an issue to a critical review and response. Those in government denied that it had any validity, since they feared the public might believe Bill Vogt and cause a fuss to interfere with plans for "growth and progress." Certainly it drew my attention to the problem of human population growth. Until then I had happily based my ideas on the expectation that the United States population would level off at a point not much higher than it had been in pre-World War II days. That had been the prevalent thinking in the 1930s, before the war. Little did I know that we had, somewhat unwittingly, already contributed a first child to what was to be known as the "baby boom." We thought it quite unfair to have our unique and priceless daughter characterized as a mere item in a bumper baby crop. But in those days we were naive, confidently waiting for life to get back to normal, meaning its prewar state. We did not know that it would never be normal again.

Vogt's book was followed by Fairfield Osborne's *Our Plundered Planet* (1949). At least that was the sequence their publication observed, though Osborne staunchly maintained that he wrote his book first. Still, the press paid far less attention to his more sober and less flamboyant style. After all, consider Vogt's ending lines: "unless, in short, man readjusts his way of living in its fullest sense—we may as well give up all hope of continuing civilized life. Like Gadarene swine we shall rush down a war-torn slope to a barbarian existence in the blackened rubble." Vogt wrote with the fervor of an Old Testament prophet. Osborne was restrained, but equally convincing. Both were well able to relate economic theory to ecological reality.

My reception of Vogt and Osborne was conditioned by Ed Graham's *Natural Principles of Land Use* (1944), Paul Sears's *Deserts on the March* (1935), G. V. Jacks and R. O. Whyte's *Vanishing Lands* (1939), and most particularly by the lectures of geographer Carl Sauer, in whose class on conservation I had been an eager student. All this was a long time before Paul Ehrlich's *Population Bomb* (1968) exploded on the public consciousness and generated far more press coverage than any of its predecessors. The timing of Ehrlich's book accounted for the difference, along with his splendid ability as a lecturer.

The world in 1948 was not in a mood to be told of limits to growth. Fresh from "winning" one world war and already moving into the not very Cold War with the Soviets, people in the United States were tired of bad news and were about to be launched into a period of rapid economic growth. They were easily convinced that the biggest problems they faced had to do with the handling of wealth and leisure, which, if they had not already arrived, were sure to come. Most Americans ignored the Korean War, even though its outcome in no way resembled a victory. By the time Ehrlich wrote in 1968, this cheerful optimism had faded. Civil rights conflicts had raged and we were well committed to an unwinnable Vietnam war that would tear the country apart. Young people in particular were ready for the news on what was wrong and how to fix it. Zero population growth appealed.

Since we had always seemed to live in some eddy of semi-poverty outside the mainstream, we were never caught up in the 1950s prosperity. My textbook *Environmental Conservation* was published in 1959. It brought us around $2,000 in royalties, enough to finance our research in southern Africa, in combination with a Fulbright grant and my university "retirement" fund. But it did not ease the economic stringency under which college professors survived. Even though it was the first ecologically oriented textbook and dealt with the problems of human population growth, in 1959 there was no environmental movement to welcome it. By contrast, in the same social environment that Ehrlich encountered in 1968, my second edition garnered over $20,000 in royalties, enough to enable us to survive in a position with the Conservation Foundation in Washington, D.C., and even to support three teenage daughters.

It has always seemed logical to me that if the population of any species continued to grow in a limited environment, the species would inevitably encounter either a shortage of resources or space or both, and that growth would be painfully brought to a halt. Consequently, it seemed logical to decide on a voluntary halt to growth before reaching such limits or shortages, and while everybody still had, or could have, enough of all necessities to live in relative comfort. I have never understood why such simple and obvious facts are not generally recognized and are even vehemently denied.

There are those who say that there are no limits or shortages in sight, and that with advances in technology, any limits that seem to be coming toward us can be overcome. Their objection has a certain degree of truth, since we have greatly increased our technological skills and have, in fact, evaded or avoided many potential limits. But the cost to the environment has been formidable. The human population of the world in Bill Vogt's time had only recently exceeded two billion. Now it nears six billion, but there has been no Gadarene-swinelike rush into the abyss, even though starvation, privation, genocide, and other slaughter are rampant in many parts of the world.

With human numbers increasing at a rate approaching one hundred million per year, I am amazed to read those who would move earth's population to space colonies. How many spaceships of what size and at what cost would it take to move one year's growth in numbers? The proposal seems absurd until you realize that it is only the elite, space oriented enthusiasts who would be transported. I would be glad to see them go. In fact, there are limits to growth, and each year the earth's productive capacity drops as desertification, soil erosion, pollution, loss of species, and atmospheric and oceanic changes accelerate.

More reasonable in their objection to population limitation

are those in the poverty-ridden regions of the earth, who see large families as a possible defense against economic and political tyranny. If you cannot trust your government, if the police and army act as your enemies, if the local community has fallen apart, whom can you trust? Your last hope is to have a lot of relatives. Even with low earning power, they are your only social security, and perhaps some may break through into economic success. The long-term future of humanity is less important than the immediate short-term survival of those you care for.

The argument that economic development must precede population limitation has a certain justification. Unfortunately, unbridled population growth limits development. Even the best managed and cared for rice paddy has limits to how much food it can produce. Population limitation and economic development must go forward together. Those who are asked to have fewer children must be able to see more than a faint hope that they and their children will survive and benefit from the changes that take place.

Leopold Kohr, a charming old Austrian anarchist, had an interesting outlook on population. He compared it to gas in a container. It is not only the number of molecules in the container that determines the pressure on its walls, but also the temperature to which the gas is heated. The more energy that goes to heat the gas, the faster the molecules move, and the more pressure on the walls. In human terms, populations such as the United States' with high energy use exert far more pressure on their environment than the more densely packed populations with low energy use, such as China's was at the time that Kohr was speaking. Since the early 1960s, of course,

China has been moving toward a high energy use status, as has much of Asia. The potential for environmental disasters thus becomes much greater.

Paul Ehrlich has carried this idea a bit further with his formula $I = P \times A \times T$, where I stands for environmental impact, P for population, A for per capita affluence, and T for damage per unit of consumption done by technology. However, he pointed out that per capita energy use could be substituted for A and T, which is pretty much where Kohr left it.

Georg Borgstrom, a Swedish food scientist based at Michigan State University, added further insight to the world population issue in 1969 by pointing out two areas usually neglected in population/resources discussions. The first has to do with domestic animals who depend in the same way as humans on the green plant productivity of the earth. To determine the total impact of a particular human population, add in the equivalent pressure on the vegetation base of their livestock. Looked at this way, Australia, a sparsely populated land because of its vast deserts, had a human population almost equivalent to that of China, since Australia had more domestic animals than people, whereas the Chinese kept few livestock or animal pets.

These relationships have been confused by the practice of relating human populations to the arable land available within a nation's boundaries. In such a relationship, Japan and the Netherlands appear to be doing an amazing job of feeding great numbers from relatively little arable land. However, Borgstrom has pointed out that both countries are dependent on a vast *ghost acreage* outside their national boundaries. This consists of *trade acreage,* the land in other countries needed to produce the food imported by these countries, and *fish acreage,* the area of ocean needed to produce the marine food brought in by their or others' fishing fleets. When these are added in, neither Japan nor the Netherlands could be seen as supporting great numbers of people from a small area of land.

In 1965 Borgstrom calculated that whereas the world human population was 3.3 billion, at that time the population of cattle alone was over 8 billion human equivalents, and for all domestic animals, plus the human population, the true feeding burden on the world's vegetation was the equivalent of 18 billion people. Today it could be twice that, or 36 billion human equivalents.

To forestall the obvious response I have often heard from vegetarian students, the answer is not to eliminate livestock so that we can produce more people, since at least a high percentage of those animals are fed from land that cannot be used to produce plant foods for people directly (arid lands, tundra, etc.) or are fed on waste products from the production or consumption of food for people. Rather, the answer is to reduce the total feeding burden on the planet and allow nonvegetarians to eat as high on the food chain as they care to bite.

Though it is true that an enormous amount of environmental damage was done when the earth's population was much lower than it is today, it is appalling to think of how much more damage would have occurred if human numbers had been greater at that time. Great areas of forests were cleared to meet the fuel and construction needs of smaller populations in the old Roman Empire. What would have happened if there had been ten times as many people armed with bulldozers and chain saws?

One of the more telling statements about the effects of overpopulation was written by Jacquetta Hawkes in *Man on Earth,* published over forty years ago:

> If the effect of numbers on the surroundings and conditions of our lives is bad, their effect on social freedom is yet more evil. Great numbers, unless they are subsistence-or-famine peasants, demand control. They are in danger like passengers in an overloaded boat, and must be shepherded, planned for, and always of necessity handled in vast groups with their impersonal, clumsy relationships. Many people who would like to fight for a reasonable anarchism are fearful to venture for fear it might lead to stampede or breakdown among the enormous urban populations where no one is able to keep himself alive if trade or services fall. Once the excuse for heavy-handed government is there, then it is seized, exaggerated, turned to a mild or cruel tyranny. (Hawkes 1953)

We have seen the inevitable increase in limits and regulations that seem to grow with each meeting of city council or other legislative body. What nobody needed to worry about fifty years ago, everybody worries about now that numbers have more than doubled. Our ancestors could throw all their wastes onto a kitchen midden without worrying about environmental effects. Even a small town cannot do this today. When there was lots of room nobody worried much about what individuals were doing. Now everyone seems to be monitoring the activities of everyone else. Big brother and sister are certainly watching you.

It is not possible to agree on what an optimum population for the world might be. My choice would be 500 million, about what it was in A.D. 1650, but I would settle for 1 billion, less than the present population of China, or of the whole world in 1850 when Charles Darwin was stirring up ideas about evolution and survival. We could accommodate 1 billion in 1,000 interconnected urban agricultural enclaves (or reservations) of 1 million each, leaving the rest of the world as a home for millions of other species, and for human wanderers. But that is just dreaming.

Bringing population growth to a halt with births and deaths in balance is difficult enough. Bringing a decrease in population is more difficult, without catastrophe, and may be impossible. With expected death rates standing as they now do at 1 percent it would take around 20 years to move the world population from 6 billion back to 4.5 billion with no births at all. It was faster climbing up the world population ladder, since birth rates have been almost triple death rates. Hence it appears that we must plan on accommodating at least 6 billion and must manage to do so without serious damage to the world's ecosystems and their wild inhabitants. Of course catastrophes could bring a decline in human numbers, and the frequency of such events appears to be on the increase.

It is easy to get totally depressed about the global population problem, but then most problems seem beyond solution when you look at them globally. The answer may perhaps be to turn the problem upside down and view it locally. After all, population growth does not occur through spontaneous generation. It involves individuals' sexual behavior, which is conditioned by their societies and environment. Growth occurs when those who are reproducing outpace those who are dying off.

Overpopulation is the overwhelming global issue. But underpopulation may seem the greater local problem. If you belong to one of the hundreds of little nations caught within the boundaries of larger nation states, it may be apparent that your people are dying off, and few are being born to replace them. Once self-sufficient tribal groups—the Lakota or the Miwok, the Yanomani or the Vedda—living as they always had lived, well within the ecological limits of their territorial environments, are being killed off by the diseases or weapons carried by invaders from outside. These are often refugees from the desperate poverty affecting the underclass in the global economic order. Can something be done to protect the endangered little nations and cultures of the planet? Yes, it could. But will it? Probably not. But don't ask me; ask the Lakota or the Miwok, the Yanomani or the Vedda, if you can find any.

Should it not be reasonable and not discriminatory to always set limits on the growth of towns and cities, planning to accommodate some reasonable level of population? Our towns and cities are made dysfunctional by excessive numbers of young being born or migrants being allowed to settle in. There would always be room for some to move in via birth or immigration as others died or moved away. But is there any imperative that says we must allow all who come along to stay? Hotels do not accept more guests when their rooms are full. Restaurants turn people away. They are not accused of discrimination for doing this. Why should cities or towns be different? Is it not our misguided belief that growth is progress and progress is good that leads to crowding? We add new suburbs, build new roads, provide access to more water, power, sewage, and so on and on, until the entire community is dysfunctional or nearly so and the quality of life deteriorates.

It has been over fifty years since I first read Bill Vogt's book. Since then I have worked with Bill at the Conservation Foundation in Washington during the foundation's halcyon days when Frank Fraser Darling, Edward Graham, Russell Train, and Fairfield Osborne were part of the team. The population question remained one of my chief interests, and from time to time I have lectured, talked, studied, and written, on and on, about the problems involved with continued population growth. It would seem that I should be tired of hitting that wall, when people seem unwilling to change their ways.

On the brighter side, however, there have been gains. Western Europe has for some time achieved a zero population growth status, and certainly its economy has hardly collapsed. China finally took a strong stand against further growth, as have many Asian nations. Yet Africa seems determined to push the limits to growth, though its balance is affected not only by the AIDS epidemic but by the massive slaughter of warfare in many countries. How unfortunate Africa would be, if the only final answer to its population problem had to come from increasing the rate of mortality.

My interest in the issues related to human population growth led me to become acquainted with Robert Cook, who was president of the Population Reference Bureau. The bureau was both a source of information on populations and an advocate for population limitation. I had subscribed to their publications before we moved to Washington. Elizabeth and I became good friends with Bob and Annabelle Cook, and I was encouraged to apply for Bob Cook's position, since he was looking forward to retiring. However, I knew my limitations and didn't see how I could take on the responsibility for the major fund raising that the bureau needed. I decided instead to stay with the Conservation Foundation.

After my experiences in Australia and New Guinea, and later with the California islands, I was not in a position to visit islands until 1967, when I was already working for the Conservation Foundation. I heard then from John Milton about the situation on the island of Dominica in the Caribbean. Dominica, partly because of the ferocious reputation of the Carib Indians, had been spared from some of the exploitation that had destroyed the forests of most West Indian islands. It still maintained an extensive area of rain forest of a different species composition from other Caribbean or mainland forests. Because of the high value attached to the wood of some of its tree species, this forest was threatened by a Canadian lumber company who wished to obtain the timber. Hoping to assess the situation for the Conservation Foundation, Elizabeth and I decided to take some vacation time and visit Dominica.

Dominica is the southernmost of the Leeward Islands in the Lesser Antilles chain that curves southward and westward from Puerto Rico. It lies between the French islands of Guadeloupe to the north and Martinique to the south. At the time of our visit, it was little disturbed by either tourists or developers. There were only two places where we could stay. One was an establishment in the foothills that seemed to have only one cabin to rent and was ill prepared for visitors. The other was an American-style beach motel. After one night at the foothill retreat where we shared our quarters with a fair number of medium-sized bats, some very large cockroaches (which terrified our daughter Lauren), and some strange sluglike creatures in the bathroom that looked like something out of *The X-Files,* we decided to try the beach location. We had first visited the island's capital, a small town named Roseau, to register our presence. We seemed to be the only tourists in the country and we attracted bunches of people wherever we went. Unfortunately, many of the women seemed to be making hex signs at us. This puzzled us since we thought we looked like an ordinary American family in an ordinary American rented car. The answer came later when we went for dinner in the motel's restaurant. None of the young women who were there to wait on tables wanted to wait on us. One of them finally came up to us. She spoke English and explained that the problem was our daughter Marlene. Marlene was a beautiful seventeen-year-old girl with long dark hair and blue eyes, but to native Dominicans she looked like the Snake Queen, a dangerous and fearsome spirit. The one waitress who was willing to risk coming near us had been educated in Barbados, a much more sophisticated area.

I should point out that the French once claimed Dominica as their own, along with Martinique and Guadeloupe. It had later come under British control, but the French influence was more impressive. Most of the inhabitants spoke a creole variety of French and not English.

Dominica is a mountainous island built by volcanic action. The mountains reach up to 4,747 feet and in many areas sheer cliffs extend to the sea; there are few beaches. The black sand beach near our motel was clean enough, but those near Roseau and the other small towns were filthy. The island's terrain featured steep slopes, heavily vegetated with trees or shrubs. All of this promised severe erosion and siltation if the timber was to be clear-cut in the usual manner that has characterized logging in tropical forests.

Obviously, those of us who visited Dominica recommended that the forests be protected through national park status or its equivalent. Removal of timber through commercial logging should be severely limited, if permitted at all. We also recommended development of ecotourism instead of exploitation of the island's resources. We noted the presence of the original inhabitants of the island (the Carib Indians) but were not able to visit their reservation or have other contact with them.

I do not know what has happened to Dominica. Following our visit the Conservation Foundation sent an expert team led by William Eddy to Dominica to prepare a report and indicate what was needed to protect the island's resources. I have heard that as a result of the international interest in the need to protect the forests, the government did hire a professional tropical forester. I have also heard of island retreats that a number of well-known rock stars from Great Britain have had built on Dominica in order to escape from the attention of the media and their overenthusiastic fans. This construction could potentially be of real benefit to the island, or equally it could become a serious problem.

14

# Uniting Nations

I had not been in Washington very long when Russ Train received a plea from Michel Batisse, the head of the science side of the UN Educational, Scientific, and Cultural Organization (UNESCO) to assign an ecologist who could write well to work with him on preparing a background paper for a conference on the rational use and conservation of the biosphere. Because I had a number of books published at that time, I was a logical candidate. In March 1966, while I was still trying to get used to living on the wrong side of the continent, I found myself on a plane heading for Tokyo and the Pacific Science Congress.

Michel Batisse was one of those people with whom I found instant rapport. Starker Leopold, Frank Fraser Darling, and Russ Train were others whom I seemed to know before we ever met. Batisse was a physicist by education, but above all he was a superb diplomat who had the ability to make things happen in the UN system. I met him for the first time in

Tokyo, along with the outstanding French ecologist François Bourlière; Ray Fosberg, the leading botanist at the Smithsonian Institute; Frank Blair, a mammologist from Texas; Max Nicholson from the U.K., head of the International Biological Programme (IBP), and others.

We discussed the need to launch an intergovernmental program to carry on where the IBP would leave off when it ended in 1968. An international scientific conference would be held in Paris in 1968, to call for and launch such a program. The Biosphere Conference, as it came to be known, would be made up by scientists from the UN member states. Such experts would not speak for their governments and hence could speak more freely. Any program proposed, however, would have to go through UNESCO's General Council (its intergovernmental governing body) for approval. My contribution was to write the background document that would ultimately be given to all participants in the conference.

However, there were to be many, many steps between the first draft I prepared and the final document. For one thing, there was considerable debate as to the meaning of the words *biosphere, ecosystem, conservation, rational use,* and *environment,* and how best to translate them into the official languages of UNESCO. Each word and sentence in the draft was picked out, shaken, and viewed from all sides. Not only did the draft have to be translated into all the languages, it could not lose its meaning in the process. Furthermore, everything had to be politically correct. The member states of UNESCO might take offense at the drop of a word or phrase. Suffice it to say that as one of the principal scribblers for the UNESCO program,

I was not necessarily responsible for the final words and turns of phrase.

The tortuous process of getting a new program established through UNESCO only became known to me through experience. I first became involved in 1966. Two years later the Biosphere Conference took place in Paris. Then a series of working group meetings took place in Paris to shape what was to become the Man and the Biosphere program (MAB). By then I was spending so much time in Paris that my daughter Marlene enrolled in the American College and we rented an apartment near the Luxembourg Gardens. It was 1971 before the UNESCO General Council approved the program and three years later before any real action began to occur. Eight years is long enough for a lot of species to become extinct, rain forest to disappear, and tribes of people to disintegrate. It would be nice to say that more than twenty years later, the program has been a great success. But it really has not accomplished much more than the IBP before it. Both programs were started with enthusiasm, but the enthusiasm was not matched with money.

A number of us had calculated during the years when the program was taking shape that it would take a commitment of $50 million to get the program off to a good start. When reality set in, the UNESCO budget allotted $500,000 to MAB, enough to keep its secretariat alive in Paris, but not much more. When MAB had developed a little momentum by 1980, Ronald Reagan was elected president and the United States pulled out of UNESCO. Since UNESCO had a full representation of dreaded, evil Communists who were allowed just as much say in its running as the saintly capitalistic Americans, the agency never won much favor among far-right conservatives. But the U.S. pullout, followed by that of Mrs. Thatcher's U.K. government, put the whole agency on short rations. It takes money to support coordinated international scientific research. Researchers must be paid salaries by someone, and their research expenses must be met. Otherwise they cannot participate regardless of their enthusiasm and ability.

It is not surprising that most people in the United States do not know much about the United Nations or what the UN is able to do. Nevertheless, the willingness to criticize and blame the UN is widespread in the population. There is no real justification for establishing the rules of the game and then blaming the players for not breaking the rules. Yet this is what we do, or rather what our governments did, in setting up the United Nations, and what we, the people, do regularly. The UN is an intergovernmental organization. The word *intergovernmental* needs stress. Representatives of the governments of the member states meet in the General Assembly, and in the various UN agencies, committees, and councils. For a program to go forward, there must be agreement, and usually consensus, among these governmental representatives. The UN usually cannot operate by simple majority vote of the representatives, since then the minority might pull up stakes and go home. If the minority included those member states who provide most of the money for the UN, then the organization would cease to function.

Observers may sometimes wonder, when the vote in the General Assembly (where all governments are represented) is perhaps 120 yes to 10 no votes, why nothing happens. But the outcome depends on who voted no. If the U.S. voted no, then it is likely that the U.K. and France did also, and there's the ball game. The General Assembly has often voted to condemn American actions. Has anything happened as a result? No.

To complicate the situation, there is the UN Security Council, established to deal with matters affecting the national security of member states. The original five members of this council, the nations that "won" World War II (Britain, France, the Soviet Union, China, U.S.A.) have veto power and can stop any action.

Having said all that, I should point out that working with the UN, or any intergovernmental organization, requires an ability to suspend your critical faculties and engage in a game of "let's pretend." You must pretend first that government representatives can actually speak for their governments. Often they do not. You must also pretend that governments will act in good faith to do the things they have agreed to do. Most often they do not. The final and most difficult suspension of your critical faculties is that you must believe that governments have the interests of their people and their countryside in mind when they vote. Sometimes they do.

Not too long ago most of the world's governments signed a "Charter for Nature." They agreed to do virtually everything that the most ardent environmentalist could ask for and also, essentially, to give up militarism and warfare. Hardly any of them meant to do most of what they agreed to do. The U.S.A., under the Reagan regime, did not bother to sign. Which was honest, at most.

If you want the UN to be more active, you must change the

rules under which it operates. But don't then criticize it for obeying your rules.

It was in 1969, when the MAB program was still being shaped, that I was asked by Batisse to work with him and Guy Gresford, an Australian who was working with the UN Economic and Social Council (ECOSOC) in New York. We prepared a paper that would be used by U Thant, then the Secretary General of the United Nations, to call for an intergovernmental conference on the human environment. This would involve my staying in New York for two weeks to prepare a draft that would then be modified by Batisse, Gresford, and others. After all had criticized it, my next obligation would be to put it in final form. The end result would be the conference that was held in Stockholm in 1972. I don't really know how much our paper influenced U Thant or even if he saw it. Eventually, I was asked if I would be willing to have it published under my name by the *Mondadori Encyclopedia* in Italy. I agreed and it was published. I do have a copy and if I could read Italian I would know what it said, but I have no English-language copies of this effort.

The Stockholm conference produced some agreements among the nation states to do the right thing by their environmental challenges. It also produced the United Nations Environmental Program (UNEP), to be based in Nairobi, Kenya.

The Stockholm conference was attended officially by heads of governments or their designates and was a formal UN conference. It was chaired by Maurice Strong from Canada. However, there was also a counterconference or forum held by the nongovernmental organizations and therefore considered more likely to bring out the truth. IUCN had a seat at the main conference and was also asked to prepare a panel discussion for the forum that was related more to its primary concerns. I was involved in putting together a panel involving a number of well-informed experts from representative countries related to IUCN. When this panel was to start, as I recall, a representative of a group headed by Barry Commoner (a professor from Washington University in St. Louis, Missouri) went on to the platform, took the microphone away from our chairperson, and announced that they would decide who the speakers and topics would be. I saw Commoner watching from the balcony as this took place. The result was a compromise with both sides represented on the panel. This was, of course, done in the hippie spirit of the early seventies—rebel, disrupt, shout down. I found myself in the awkward position of agreeing with much of what the rebels wanted to accomplish while disagreeing with their methods for achieving their goals.

In view of the Stockholm conference, UNESCO was interested, mostly through the influence of Michel Batisse, in publishing a book that might make known their role in the program of Man and the Biosphere. I was asked to write this and agreed. It gave me the opportunity to work once more with my former editor Peter Ritner, who had left Macmillan and joined World Publishers. The result, *Planet in Peril?* was jointly published by World, UNESCO, and Penguin in 1972.

By the time of the Stockholm conference I had left the Conservation Foundation and was working for IUCN in Switzerland. How and why this happened requires some explanation.

During our 1963 trip to Costa Rica, and two years later in the Airlie House conference on future environments, I had become acquainted with Gerardo Budowski, a tropical ecologist whose ideas and facility to express them in a number of languages impressed me very favorably. When it turned out that he was coming to Paris in 1969 to be in charge of the MAB program, working with Batisse, I was pleased. Then it developed that IUCN was looking for a director general. I was one of those who recommended Budowski for the position. In due time he asked me to join IUCN as senior ecologist, working with him to develop and support research related to nature conservation. This job would involve another complete change in the direction of our family life. We had worked with the Conservation Foundation and through it UNESCO for four years and had bought a house in Washington. Moving from there to a site on Lake Geneva in 1970 was a major undertaking. Sandra was still in Berkeley and Marlene had decided to major in biology at the University of Colorado in Boulder. We arranged for Lauren to go to a prestigious girls' boarding school, Foxcroft, in Virginia. It looked as though Elizabeth and I would be "alone at last" in Switzerland. We decided to rent out our Washington house. It was necessary for me to go on ahead to start work with IUCN while Elizabeth stayed until everyone was settled and the house was rented. I found an apartment for the two of us in Préveranges, a small village near Morges. When Elizabeth arrived, we found a more spacious flat on the sixteenth floor of the only high-rise in Morges, with a view of the Alps beyond Montreux to Mont Blanc and beyond. Lake Geneva with its sailboats and ferries was in full view. It was a five-minute walk from the IUCN and World Wildlife Fund offices. Could anything be more ideal?

In retrospect the year 1969 seems to be excessively crowded with travel and other activities. In that summer Elizabeth and

I spent a month in Rome, where I acted as a consultant to the forestry and wildlife section of FAO. This happened because my old friend from deer-study days and work in Southern Rhodesia, Thane Riney, had joined FAO and invited us to spend some time with him and his wife, Anne. Another friend, the Dutch ecologist Tony de Vos, was also working with FAO and offered us the use of his flat in Rome for a month while he was away. Our older daughters, Sandra and Marlene, would be on a work-exchange program for the summer and would be in London. We signed the youngest daughter, Lauren, up for a summer program in a school in Neuchâtel in Switzerland. So the two of us were free, more or less. When Lauren's school program ended, we brought her to Italy and she and Elizabeth visited Naples and Capri. We also traveled around central Italy with Thane and Anne. I can't remember doing any significant work, but we did have a lot of fun.

15

# Return to Africa

Though I did not get back to Zimbabwe despite several opportunities, my work in Southern Rhodesia seemed to have brought me a number of enthusiastic friends in South Africa. The National Parks Board in South Africa and the Rhodes University at Grahamstown invited me to address the South African Association for the Advancement of Science and to visit Kruger National Park, with all expenses paid for Elizabeth and me. So once again, in 1974, we were on a plane across the Mediterranean and the Sahara for a much more enjoyable flight than the one in 1959. We did not stop at Khartoum, but went on to Nairobi, where we changed planes, and then to Johannesburg, where we received VIP treatment all the way.

At Kruger National Park we were taken to witness and participate in many park activities. They included a helicopter flight intended to eliminate some of the excessive population of buffalo that was believed to be competing with and reducing the numbers of other grazing species. Both buffalo and elephants indicated their hatred for our copter. One huge buffalo stood on his hind legs and pawed the air to show what would happen if he could only get airborne. A game ranger with a rifle fired immobilizing darts from the copter. When the darted animals collapsed, game scouts rushed in to kill them. From there, their bodies were trucked to the park abattoir, where they were converted into canned meat for the European market. An ironic development from our game ranching efforts fourteen years earlier. Our sympathies were on the side of the buffalo. We did, however, have to admire our Belgian pilot, who did an amazing job of twisting and turning the chopper in pursuit of the fleeing herds.

For me the only really scary event during our visit came when Elizabeth and I went out with an American researcher who was studying leopard populations, Jim Bailey, from the University of Idaho. He had established cage traps for the big cats, baited with the kind of meat leopards love to eat. When he caught a leopard he would immobilize it, put a radio collar around its neck, take various measurements, and then, when it had recovered sufficiently from the drug, release it. The leopard's travels could then be followed as it went about its daily routines.

When we arrived at the trap, we did find a very angry male leopard. I watched while the cat was immobilized, weighed, measured, and collared. The trap was located on the bank of a dry stream. While we waited for the leopard to recover and be released, the sun had set and it was beginning to get dark. I realized that Elizabeth had gone down into the dry streambed with the idea of photographing the leopard as it took off from the trap. I went to join her and recommend that she get back closer to the Land Rover in the event that the leopard was still in a feisty mood. However, the cat was not interested in fighting, only in getting away. It shot out of the trap, crossed the stream, and disappeared into the woods on the other side.

Meanwhile, however, I found Elizabeth virtually surrounded by a number of spotted hyenas. She had been concentrating on the leopard and her camera and had not noticed the hyenas' arrival. We both made a cautious retreat back to the Land Rover, but the hyenas only watched us leave and did not show any aggression. I suspect they had been attracted by the meat in the trap and were only waiting for us to go away. It was too late for photographs, but the light from my flashlight picked out the gleam of the animals' eyes. I found it unnerving, but Elizabeth was not disturbed at all, only frustrated that she did not get her pictures.

Of course, in elephant country there are always little surprises, such as encountering an enormous bull elephant in a bad mood when there is no way of getting off the dirt road and you are driving a little Volkswagen beetle. You learn to drive in reverse at high speed. There was also a light-hearted occasion that Elizabeth immortalized in a painting. We encountered four young giraffe who were playing games—chasing one another and running around in a big circle!

We also had occasion to talk with the prime minister, Vorster, about the importance of South Africa's national parks in maintaining a good image abroad, despite its apartheid policies. The Kruger National Park, in the Transvaal of South Africa, was one of the first national parks anywhere. Many more national parks have since been added and they have survived all the political and racial turmoil. Unfortunately, the one aspect of the wild scene that has been lost may be the most important, the hunter-gatherer people, the Bushmen, who lived for centuries—if not millennia—in balance with the wildlife, doing no harm.

16

# Ecosystem and Biosphere People

During December 1974 I was invited to a symposium to be held at Cambridge University in England. It focused on the future of traditional "primitive" societies and brought together anthropologists, ecologists, and people from many disciplines with similar interests. Sir Edmund Leach, professor and provost of Kings College, Cambridge, was the primary organizer. I was asked to prepare and present a paper on difficult marginal environments and the traditional societies that exploit them. This I did, and the results were to be published in whole or part in many other books and journals. The part of my speech that had wide appeal was my classification of the world's people into two categories, *ecosystem people* and *biosphere people.*

> The ecosystems within which traditional "primitive" societies have been able to survive are not necessarily those that are particularly well suited to the ways of life prac-

ticed by the members of these societies. Rather they are ecosystems with refuge qualities that have discouraged their invasion by members of the globally dominant cultures. Although these ecosystems represent a wide variety of ecological conditions, they have in common the quality of being located at extremes of global temperatures and moisture gradients, or of occupying the most difficult terrain in a physiographic sense. By contrast, the middle ground of moderate temperature, moisture, or relief has been taken over by the dominant cultures.

Traditional "primitive" societies are ecosystem-dependent, meaning that they occur within a single ecosystem, or at most make use of a few ecosystems, and are subject to the ecological controls within an ecosystem. By contrast, the globally dominant cultures draw upon the resources of the entire biosphere and can override the normal controls within any single ecosystem. This makes possible a much more complete disruption or destruction of the components of an ecosystem than is possible to an ecosystem-dependent society. Although indigenous cultures have developed in such a way as to provide for an ecological balance between people and their environment, this balance is upset by interference from biosphere-dependent cultures. When this takes place, an increase in the populations of traditional societies may occur, or they may gain access to technologies not previously available, which in either case leads to disruption of the systems within which they had previously coexisted.

All of the ecosystems occupied today by traditional "primitive" societies are particularly vulnerable in the face of the major modifications brought about by the globally dominant cultures. Although the arctic tundra is proba-

> bly the most fragile, even the supposed stability and resilience of the humid tropical forest biome is more apparent than real. The inability of tropical rain forest to recover completely from massive disturbance has led to its being characterized as a nonrenewable resource.
>
> I believe it is useful to recall that the ecosystems now occupied by indigenous people do not necessarily represent areas that are well suited to the economies and ways of life characteristic of their societies. Instead, these ecosystems are refuges that have qualities which have hitherto made them resistant to invasion by the dominant cultures of the earth. Since the dominant peoples have lacked interest in or have been unable to cope with these lands, they have been left to those who have learned to survive there. Thus, the Bushmen of southern Africa once were to be found throughout many of the ecosystems of that region. They survive in the Kalahari in areas that have been considered essentially worthless by the dominant, recent invaders of that sub-continent. The Australian aborigines, who once occupied much of the continent, now live in the deserts or desert-fringes that until recently were of no interest to the dominant culture. (Dasmann 1975)

In February 1975, two months after the Cambridge symposium, I addressed the South Pacific Conference on National Parks and Reserves in Wellington, New Zealand, and presented the ecosystem and biosphere concept to the Maoris and Europeans who were present. It seemed to be well received, but the effect of a speech on paper is difficult to evaluate until you see who picks it up and uses it.

Those who have grown up in Europe or North America, and have assimilated the view of history proclaimed in those civilizations, know that once there was a paradise on earth and it was called the South Pacific. For more than two centuries, adherents of Western technological culture have been fleeing from the supposed benefits of their culture in search of that paradise or its remnants. The more they have searched, the farther it has receded from their vision. Finally, in desperation, they have attempted to recreate it in the tourist lands of Hawaii or Tahiti. But the new model has not been pleasing to the soul.

More than a century ago, the European invaders of North America were pushing into what they called the wilderness of the West. Most were concerned only with the problems and perils of each day, but a few could see the realities about them with a vision that transcended the purely utilitarian. The artist George Catlin was one of these who was greatly disturbed by the destruction of the North American bison and its consequences for the future of the Plains Indian people. He had a proposal that he hoped might save both wildlife and people:

> And what a splendid contemplation too, when one . . . imagines them as they might in the future be seen . . . preserved in their pristine beauty and wildness, in a *magnificent park,* where the world could see for ages to come, the native Indian in his classic attire, galloping his wild horse, with sinewy bow, and shield and lance, amid the fleeting herds of elks and buffaloes. What a beautiful and thrilling specimen for America to preserve and hold up to the view of her refined citizens and the world, in future ages! A *nation's Park,* contain-

> ing man and beast, in all the wild and freshness of their nature's beauty! (quoted in Nash 1967)

This proposal, made in 1832, is commonly regarded as being the first request that a large area of wild America be set aside as a national park. Let us ignore for the moment the obvious chauvinism, since this characterized most nineteenth-century Europeans. Catlin's was no modest proposal, for he wanted the entire Great Plains from Mexico to Canada set aside for the protection and use of those people and animals to whom it rightfully belonged.

At the time there was no receptive audience. The West was being won by those to whom, in Catlin's words, "*power* is *right* and *voracity* a *virtue*." Viewed from the other side, however, the howling wilderness that these narrow men were trying to subdue looked quite different. In the words of Chief Standing Bear of the Oglala Sioux Indians:

> We did not think of the great open plains, the beautiful rolling hills, and winding streams with tangled growth as "wild." Only to the white man was nature a "wilderness" and only to him was the land "infested" with "wild" animals and "savage" people. To us it was tame. Earth was bountiful and we were surrounded with the blessings of the Great Mystery. Not until the hairy man from the east came and with brutal frenzy heaped injustices upon us and the families we loved was it "wild" for us. When the very animals of the forest began fleeing from his approach, then it was for us the "Wild West" began. (quoted in McLuhan 1971)

Forty years after Catlin's time, in 1872, the Congress of the United States proclaimed the world's first national park in the Yellowstone region of the territory of Wyoming. Eighteen years after that, shortly after Christmas in 1890, the United States Army surrounded and massacred most of the last independent band of the Sioux Indians at a place called Wounded Knee in South Dakota. A survivor, Black Elk, stated: "something else died there in the bloody mud, and was buried in the blizzard. A people's dream died there. It was a beautiful dream . . . the nation's hoop is broken and scattered. There is no center any longer, and the sacred tree is dead" (quoted in Brown 1971).

Later, half of Catlin's dream was realized. The animals were given the first national park. The Indians had a different appointment with destiny.

The national park movement, started at Yellowstone, has been generally regarded as a great success. When many of us met at Grand Teton and Yellowstone, in 1972, we noted that there were more than 1,200 national parks or their equivalents throughout the world that met the high standards of the United Nations' list (IUCN 1974). The centennial of the national parks movement was a stirring occasion for those of us who favor conservation. But I wonder how many noted that in the following year, 1973, a band of Oglala Sioux and other Indians seized and held the town of Wounded Knee, South Dakota, for many months, in a dramatic protest against treaty violations. They were asking, among other things, that the lands that had been guaranteed to them by solemn treaties with the United States government, be in fact given back to them. Some of these lands are in national parks (Burnette and Koster 1974).

As Tururin, chief of the Pataxo Indians of Brazil, has put it, "We Indians are like plants: when changed

> from one place to another we don't die but we never fully recover. We will not leave here because even before the reservation existed we already lived on this land. It may be bad, it may be good but it's our land."
> But in Brazil the previously isolated ecosystem people are being threatened or destroyed by the massive drive for the exploitation of Amazonia—a process far less benign than any effort to create national parks.
>
> It is characteristic of wealthier biosphere people that they do not want to stay at home. They wander the globe always searching—searching for something they seem to have lost along the way in their rush to capture the resources of the world and accumulate its wealth. Thus they give rise to the tourist industry, and this in turn provides a financial justification for creating and maintaining national parks. In these parks the wanderers can see some of the wonders that they left behind, and can pretend for a while that they have not really destroyed the natural world—at least not all of it. They will pay highly for this experience. But for some reason the money nearly always tends to be channeled back into the biosphere network. It does not go to those who were once ecosystem people and who have the strange idea that what is now called a national park is really just the land that was home. (Dasmann 1976)

Over time the original meaning of my terms became somewhat scrambled and some assumed biosphere people were those who could draw on the resources of the biosphere but, like ecosystem people, did not exceed the limits of sustainability. I agree that we could reach this goal if we all learned to live at sustainable levels of resource and energy use and did not also threaten the survival of other species or the ecosystems in which they live.

In 1988 I wrote my latest version of the ecosystem-biosphere-people issue that I now present again here.

> Some years ago I found myself guilty of categorizing people into two groups. The occasion was a meeting of anthropologists and ecologists held at Cambridge University, where I was asked to consider the ecological relationships of "primitive people," meaning hunter-gatherers, hunter-gardeners, nomadic pastoralists, and others who had retained a self-sufficient existence, largely cut off from the influence of the dominant cultures of the earth. What categorized these people, from one viewpoint, was their ability to survive, year after year and often for centuries or longer, on the resources of a single ecosystem or contiguous and related ecosystems. Through one means or another, they had achieved a sustainable way of life that did not bring about changes in the natural biota deleterious to their continued existence. I called these "ecosystem people."
>
> By contrast, apparently since the rise of civilization, the societies and cultures that are featured in the history books have demonstrated a capacity to overexploit and convert productive ecosystems into near barren wastelands. Although this capacity has in the past contributed to the collapse of kingdoms and empires, the practice of overexploitation (or the consumption of natural resource capital) continues today. Now those who are tied in with the global economy can continue to survive despite destructive exploitation of local environments. Through the

exchange of the products of industry they can draw on the total resources of the biosphere to override the controls previously exercised through the productive limits of local animal and plant populations. One needs only to look at the source of origin of most the foods consumed by city populations in the industrialized world to visualize the trade routes that pour resources from throughout the biosphere into their hands. I have used the term "biosphere people" to categorize those who are linked to and dependent on the continued functioning of global trade in these resources. I now regard the term as unfortunate since it suggests that we of the global economy have achieved the same sort of symbiosis with the total biosphere that characterizes ecosystem people and their home environments.

Obviously, this categorization is oversimplified. Certainly most people in the world are neither one nor the other, but caught somewhere in between these two positions. Since the contact between the biosphere cultures and ecosystem people is usually destructive to the ways of life of the latter, many of these have lost their ability to survive as they had in past times, but usually have gained little of the security that for the more fortunate derives from being part of the global economy.

It is also true enough that there have been tribal people who did overexploit local resources. Paul Martin (1973) has pointed out the apparent role that early North Americans played in exterminating the megafauna of this continent. More certain is the role of ecosystem people in bringing about extinction for many species that once occupied islands such as New Zealand, Hawaii, and Madagascar. I have suggested, however, that a distinction must be made between invaders and natives. When humans occupy new and unfamiliar ecosystems, their effect on the biota is likely to be devastating initially. As they settle down, learn the ecological realities of their new home, and come to terms with its limits, this destructive impact is mitigated until they achieve the level of symbiosis with their environment that characterizes long-settled natives.

The invader effect, however, can be renewed when ecosystem people take up new technologies—particularly where these are brought from outside. An obvious example involved the introduction of the horse into Plains Indian culture, which brought a new and potentially destructive relationship with the bison herds. Whether or not a new balance was being reached can no longer be answered since both Plains Indians and bison were decimated by the new wave of exotic invaders, the Europeans. Similarly, the introduction of a new trapping technology to the Indians of northeastern America combined with a shattering of their religious contract with wild animals to bring a devastating impact upon the wildlife of that region (Calvin Martin 1978). (Dasmann 1988)

17

# The Edges of the Sea

Though IUCN and the World Wildlife Fund supported many excellent research projects and conservation programs around the world, most if not all of these were directed toward the protection or management of terrestrial species and communities. Yet most of the planet's surface is water, not dry land. During my time at IUCN I was reminded often by my marine biologist friends, Carleton and Jerry Ray, that oceans and seas occupied 70 percent of the globe. And I believe astronauts were the ones who first recognized that our earth is the blue planet, since viewed from outer space it is a glowing sapphire, a living gem in an otherwise lifeless solar system. Yet the oceans are under attack from the same forces that disturb the continents and islands.

Except for near-shore waters, the oceans have no owners. Whereas on dry land nations go to war over ownership of barren mountain peaks or desert islands, the seas and oceans are for the most part unclaimed. This fact makes the conservation of marine environments and species much more difficult. Thus, for example, the United States could pass a law protecting dolphins, but the law could not stop the fishing fleets from Japan or Russia from catching them. Several meetings were held on the island of Malta during the sixties, to try to get agreement on an ocean regime that would involve an international authority to protect marine environments. I remember one in particular where I advised David Brower, then the president of Friends of the Earth, not to focus conservation attention on whales because, I said, nobody can identify with or feel friendly toward whales; it would be like feeling affectionate toward submarines. Choose something smaller and furry. Fortunately, he ignored my advice and helped launch a highly effective whale conservation program that aroused worldwide sympathy toward the plight of endangered whale species.

IUCN took some steps toward including marine conservation projects in its programs by establishing a marine committee in the Commission on Ecology. However, its efforts did not seem to have accomplished very much by the time I became involved. Working with Carleton Ray, Sidney Holt of the FAO, and others, we had a meeting in Rome that led to the reestablishment of the marine committee. We were able to involve a reasonably representative group of marine scientists to assist the committee in determining priorities for projects. The concepts presented by Carleton on determination of critical marine habitats were accepted as a basis for establishing priorities, along with consideration of endangered species.

Critical marine habitats are those on which endangered

species depend for their survival along with those on which a large number of species depend, whether or not they are endangered. Within the first category are the bays and estuaries of Baja California, where the endangered gray whales go to breed. The second includes coral reefs, sea grass flats, mangrove swamps, and perhaps oceanic trenches, with their many strange life-forms. Estuaries where freshwater from rivers and streams mixes with saltwater from the seas are homes for many species adapted to the lower and changing salinity of estuarine waters, along with other species that spend part of their life cycle in seas or oceans but also depend on estuarine environments at some stage of their development. And yet critical marine habitats are often the areas subject to the greatest pressure from on-shore developments including pollution from human wastes, dredging for harbors, and filling in to provide more space for housing, transportation or other commercial or industrial uses. On top of these, overfishing exerts still greater pressure on those marine species that occupy the edges of the sea or depend on it for their populations to thrive.

The ratio of edges to interiors is of great interest to conservation biologists. It may have been Aldo Leopold who first pointed out the importance of edges for game species, noting that many, if not most, game species either require, or are benefited by, the edges between plant communities. Ring-necked pheasants, for example, or quail, or cottontail, do not survive in grain fields even though they will feed on grain. But add a hedgerow or a brush patch next to the grain field and you find that many game species will make a home there. Wildlife diversity is favored by vegetation diversity, and so too is the abundance of animal life. That has certainly been borne out by our work on deer, which are to a great extent dependent on edges.

As coral reefs illustrate most vividly, the edges of the oceans support the greatest diversity of animal life. Looking only at an atoll's terrestrial area, you might say that a coral atoll is all edge and no interior, and the same applies to small islands regardless of their origin. But when too much edge prevails, an environment loses species that are dependent on the interior of vegetation. And some marine species require edges in their juvenile stages though they may as adults occupy the open ocean. Obviously, gray whales are at home in the open ocean, but they go to the sheltered bays of Baja California to give birth to their young. Salmon and steelhead trout spend most of their lives in the open ocean but return to lay eggs in freshwater streams, where the young fish grow and survive. There are many species with similar life histories requiring different habitats at different phases of their life cycles. One of the strangest of these is the marbled murrelet, which spends most of its life on the open sea but moves into the tops of old-growth redwood trees to nest and raise its young.

It would be encouraging to say that the efforts of the IUCN committee led to increased funding for marine research and conservation, but I can't really say that it did. We had a few meetings, notably one on the French island of Guadeloupe, neighbor of Dominica in the West Indies, where we began to establish goals and priorities, and later a meeting in Bergen, Norway, and a conference on marine national parks in Tokyo. Sidney Holt played a prominent role in the International Whaling Commission, which led to increased protection for the various species of whales.

I suppose I helped out a bit when I was appointed to the California Fish and Game Commission by Governor Jerry Brown in 1978, but my contribution was small and I was more involved with condor conservation than with marine affairs.

Oddly enough, it turned out that Sidney Holt and I had both applied for the professorial position in environmental studies at the University of California at Santa Cruz in spring 1977. As the slow work of the search committee proceeded, we both ended up in the top three candidates, and soon it became obvious that we were the two leading candidates. We became aware of this fact when we were holding the first IUCN Marine Committee meetings at UCSC. Sidney and I conferred one evening at a party and decided that neither of us really wanted a full-time position, so we would each take a half-time salary and teach alternating quarters at UCSC. Sidney would start with the fall quarter 1977 and then go back to FAO in Rome, while I would start with the winter quarter, he would come back for the spring quarter 1978, and I would take the fall quarter 1978. We made this proposal to Ken Norris, who held the chair of environmental studies and also happened to be a world renowned expert on whales and dolphins. Though our proposal was unprecedented, Ken managed to convince the administration it was a good idea. So we were appointed, and Elizabeth and I moved to Santa Cruz.

Sidney made a great impression on the students and started them going on a number of projects that I inherited when the winter quarter started. However, when the spring quarter arrived, Sidney did not; he was too much involved with his other jobs. In fact, he never returned to pick up his half-time position and I finally had to go on a full-time appointment. I don't know why it turned out that way even though I talked with Sidney in Guadeloupe and Bergen.

My interest in marine environments paralleled my interest in islands, which began with my World War II experiences on the near-shore islands of Australia and the big island of New Guinea with its associated smaller islands. I suppose one reason for my attraction to islands is the expectation that I will find something new and different when I visit them. I usually do. Islands are centers of endemism. The larger the island and the more isolated, the greater the likelihood it will be inhabited by new and different varieties of plants and animals.

> We are all island people. Most of us did not fully realize this until we sent our astronauts into the skies and were shown a picture of island earth against its background of a black and star-specked sea of space. We know now that we live on a small area of habitable land in an inhospitable ocean. There is no continent to which we can flee, only other island planets where the environments are hostile and forbidding. We must do the best we can with what we have—other resources are not available. If we destroy those elements that we need for survival, we cease to exist.

The comments above come from a keynote address I presented to the Regional Symposium on Conservation of Nature held at Noumea, New Caledonia, in August 1971. New Caledonia is one of France's Pacific possessions and in economic terms one of its most valuable, since it has rich deposits of nickel ore that can be obtained by strip mining. The French government is inclined to be highly sensitive to criticism of its governance or management of these islands, and most New Caledonians support strong independence movements. It is not surprising therefore that I caused more than a little controversy with the comments I made in my address when I focused in on New Caledonia.

> I am not an expert on New Caledonia. Much of what I thought I knew about the island before I came here was either wrong or out of date. However, a good medical doctor does not need exhaustive tests to say that a patient is sick, and an ecologist does not need exhaustive studies to say that an island is sick. New Caledonia is an island that is ecologically sick.
>
> I wonder if there are many other places where so much damage has been done by so few for so little purpose. Of course there are, the same things are happening elsewhere. Indeed, the first law of the environment today seems to be—"no matter how bad you think things are—the total reality is much worse."
>
> But it is disturbing to see the processes of destruction so actively at work as they are here. The unique forests of New Caledonia—did they really have to be destroyed? Was the profit to be obtained by their destruction so high as to justify the consequences—the scrub-covered hills, the barren, eroding slopes? Admittedly, you have the tattered remnants of the old forests, and these are now completely priceless. Not only do they serve to remind one of what the island must have been but they are the means by which the island might some day be restored.
>
> One wonders—do the powers that be, the public and private authorities, here and in France, really intend to

> destroy this island completely? To totally strip the surface of the land and leave nothing but devastation behind? Surely nickel or iron ore are not so valuable as to justify such a course. People must continue to live here. They cannot all go away. Perhaps those who make the highest profits should be required to live on the most devastated lands.

Well, Michel Batisse warned me that the French were very upset by my remarks. After the formal meeting was ended, the governor of New Caledonia held a reception for the conference participants. Though I stood in the receiving line, the governor turned away and did not shake my hand or greet me. The next morning, the local newspaper described the conference under the heading "Swiss go home!" Fortunately they did not know I was American.

Strangely enough I lost my copy of the address in Noumea. I had not stuck to my prepared text and did not know what I had said that was so offensive to the New Caledonians. Now, on rereading the transcript of my talk, I can understand the reaction. It was bad form to say what I did that day in August 1971. Admittedly, what I said was true enough, but the New Caledonians who were at the symposium were already aware of the island's deterioration, and some of them, at least, were trying to do something about it. My remarks could have helped but instead turned out to be insulting. I will try to do better in the future.

I went from Noumea to Sydney, Australia, to attend and speak at the twelfth Pacific Science Congress, held in Canberra. My subject was reconciling conservation and development in the coastal zone. I don't think I insulted anyone there and I may even have been helpful.

Islands are characterized by limited space and often fragile ecosystems. They are vulnerable to the kind of pressure that can be brought to bear on them by the workings of the global economy and its technology. The special characteristics of islands have resulted in their receiving particular attention in the UNESCO Man and the Biosphere program. Of the thirteen project areas comprising the original MAB program, one was devoted to the ecology and economy of islands. I have not been involved personally in this project nor have I heard about what it has accomplished, but its usefulness should be apparent.

18

# The Incident in Kinshasha

For seven years I worked for the IUCN, based at that time in Morges, Switzerland. The acronym is apparently meaningless to most Americans but stands, as I noted, for an organization then known as the International Union for the Conservation of Nature and Natural Resources, and now known simply as the World Conservation Union, but retaining the logo IUCN.

My career at IUCN was in the role of senior ecologist, the number three position in the IUCN secretariat, under the director general and deputy director general. When I arrived in Morges in 1970 almost the entire staff was new and we were all trying to build a new organization, on the basis of a large grant from the Ford Foundation.

IUCN may well have been the most complex international organization ever devised. Its membership included all the UN's governmental organizations and nongovernmental organizations (NGOs), among them the most cantankerous environmental organizations on earth. In theory its governing body, a general assembly, met every three years. The group was made up of delegates from all the member organizations, including both governmental agencies and private groups such as the Sierra Club and Friends of the Earth–U.K. It was essentially a bicameral assembly, since everything had to be approved by both the governmental and nongovernmental members. Thus if the Canadian NGOs wanted to halt the destruction of Canada's coastal forests or prevent the James Bay Development, the Canadian government's delegates could block their resolution on the issue. If this meant that the final approved program was somewhat bland, so be it.

During the three years between general assemblies the organization was governed by an executive board elected by the general assembly, which met as needed during this period. In fact, the board usually approved the recommendations of the director general, since the members could hardly keep up with the IUCN secretariat's daily work.

To add further complications, IUCN at that time had commissions, made up of experts in species survival, protected areas, ecology, education, law, and environmental planning. The commission chairpersons sat on the executive board and held considerable ability to influence decisions. To make matters still more complicated, IUCN was tied to the World Wildlife Fund, which provided a good share of its financing and tried to influence its programs and projects. IUCN was also tied to the UN Environmental Program (UNEP), based in Nairobi, which supported some major activities, and to UNESCO and FAO, which also provided support for some projects.

In the face of all this it is quite amazing that the IUCN secretariat and commissions ever managed to get any real work done. Paperwork, intended to keep all the players satisfied, was voluminous, but actually quite a bit of real conservation work in countries around the world was accomplished.

Among the many problems IUCN faced was in the field of personnel. There it inherited the UN problem, that the paid staff must be somewhat representative of the membership. That meant that you could not necessarily hire the best, most qualified person to do a particular job, but had to maintain a balance among Africans, Asians, Europeans, North Americans, Latin Americans, and so forth, as well as take into account the socialist and capitalist divisions of the membership. Hence we sometimes found ourselves with less than qualified staff.

When I was with IUCN the director general was Gerardo Budowski, the brilliant plant ecologist and expert on tropical forests who had moved from UNESCO to take over as chief executive of IUCN. I had followed him there. The deputy director general, Frank Nichols, had come to IUCN from Thailand but was an Australian citizen. His background was in physics but he was primarily an administrator. Unfortunately, he conveyed the impression, perhaps a false impression, that his interest in conservation was minimal and that he would be equally happy working for the exploiters. My role in the secretariat was at first heavily involved with IUCN policy and programs, so that the three of us formed what the Russian board members called a troika, based on our behavior at council meetings. However, I was to move away from policy and administration to concentrate more on research.

Anyone might assume that an organization directed toward the highest human ideals of nature conservation and headquartered in a delightful old chateau by the shores of Lac Leman in the quiet town of Morges, not far from Geneva, would bring out the very best in its staff. It was an idyllic setting in which idealism should have thrived, but it did not.

I became heavily involved in a number of projects, such as mapping the world's biotic areas in order to give us a firm base for directing expenditures on research and conservation in third-world countries. Regrettably, the old IUCN along with the World Wildlife Fund had directed projects toward the areas of interest to its board and leaders, or folks with money, rather than the areas of greatest need. There were or had been numbers of projects in East Africa, because the British VIPs liked to go there. There were few in West Africa. There were many projects in India, another favorite hang-out of wealthy British, but few in the rest of Asia except for Indonesia, which the Dutch VIPs favored.

In working with UNESCO and later with the IUCN in Switzerland, I found that both organizations tended to use the biome system of classification of natural communities originally developed by Frederick Clements and Victor Shelford (as I mentioned in chapter 7). However, this system makes use of vegetation structure without paying great attention to the species involved except for the dominant plants. It could be considered the opposite of the system developed by Braun-Blanquet at Montpellier in France, which pays attention to the species without emphasis on structural arrangements of the plants. The biome system (or vegetation formation) was further developed and expanded by Ray Fosberg, botanist at the Smithsonian Institution, as the classification used primarily by the IBP since it had been developed to cover the world's vegetation. Fosberg did not pay attention to the successional status of the vegetation, unlike Clements, but described what was growing on the ground, not what it might over time develop into as a climax that would then hold the ground for a long time, perhaps thousands of years in the case of a redwood or big tree *(Sequoiadendron)* forest.

The question that I felt was of first importance for IUCN was what was the purpose of the classification; for what would it be used? The answer, as I saw it, was for species conservation as a primary objective for IUCN. To protect species, however, you must protect their habitat, the biotic community of which they form a part. Thus it was necessary to protect not just one tropical lowland rain forest in the Amazon, but all the different kinds of tropical lowland rain forests. All might have a similar vegetation structure and support ecologically equivalent animal species. But a leopard is not a jaguar, even though they both occupy similar niches and both have the same status as top predators in their food chains.

If IUCN, the World Wildlife Fund, or other conservation agencies were to put most of their money and energy into rainforest conservation in the Amazon, they could lose hundreds or even millions of species found only in Asian, Australian, or African rain forests. A useful classification would therefore emphasize species composition as well as vegetation structure. Such a system had been developed by Lee Dice in Michigan and was called biotic provinces. Though Dice was concerned only with North America, I decided to apply it worldwide. To help with this I was able to draw on the knowledge of members of IUCN's ecology commission and species survival commission, as well as others. With this, I drew the boundaries of the major biotic provinces based on vegetation maps, floristic provinces, and bird and mammal distribution.

These attempts to define and classify the floral and faunal areas of the world are based on work done long ago. In 1849 Alexander von Humboldt noticed that changes occurring in vegetation at increasing elevation in mountains resembled the changes he observed traveling from the tropics toward the polar regions. And in 1876 Alfred Russell Wallace noted the major faunal and vegetational regions of the world, which were to become known as Wallace's realms. Before him, P. L. Sclater tried in 1858 to classify the natural regions of the world on the basis of their bird fauna. Thus early researchers had seen that the Australian fauna was distinctly different from that of Asia, since the continents had long been separated. Similarly, the fauna in North America was distinctly different from that of South America: few North American mammals had crossed over the land bridge of Panama that emerged from the sea relatively recently (in geological terms). Deer and mountain lions had moved south across the bridge; tapirs and peccaries had moved north. But the major faunal differences between the continents remained. The Asian tropics had more in common with Africa than with Asia north of the Himalayan barrier.

Dice's biotic provinces and similar work defining floristic provinces were on a much smaller scale than the differences in Wallace's realms. My first task at IUCN involved pulling together all this material in an attempt to define and map the biotic provinces of the world. Some of my colleagues were reluctant to draw lines on a map, but if I drew the lines they often responded quickly, pointing out my errors. I had trouble getting any response from researchers in China and the U.S.S.R. and was struggling to define the biotic provinces of Latin America. As I got ready to produce a final map I turned to Miklos Udvardy at Sacramento State University. The textbook of zoogeography he had produced seemed to have a lot in common with what I was doing and he was willing to work with me. Fortunately he was also much more familiar than I with Latin America and Central Europe. John King, my old pal and fellow sufferer from the 32d Infantry Division who was by then working for the CIA, finally got me the information on China. The end result of all this was presented as an IUCN occasional paper (Udvardy 1975) and then as an excellent map and accompanying text by UNESCO. For reasons now obscure to me, I did not put my name on the final product and so missed out on any fame that might have adhered to it. Udvardy was certainly willing to have me as coauthor or senior author, but I declined. Does it matter? No.

My work with biotic provinces led to my becoming involved with the concept of bioregions (though I preferred the label "ecocultural regions," since far more than biology goes into defining their boundaries). Teddy Goldsmith, editor and publisher of *The Ecologist,* a British journal, sent me a copy of a paper submitted to him by Peter Berg. He asked that I read it and, if I thought it to be basically worthwhile, to work with Berg to rewrite it so it would be publishable. I did this and the result was an article, "Reinhabiting California," by Berg and Dasmann (1976). This struck a responsive chord among the countercultural people of that time and it was reprinted in a number of books and journals. Essentially, it pointed to the need for a new political economy based on the ecological characteristics of an area and the human cultural activities that could lead to sustainable uses of that place. Instead of a map showing only floral, vegetational, and faunal provinces, it would also take into account human perception of the area they considered to be home. It called upon people to become well acquainted with the area in which they lived and to know where their water, food, atmosphere, energy resources, and other necessities for life came from and what was happening to them. It asked that people develop a reasonable degree of self-reliance, or at least the knowledge and ability to survive in the place where they lived. In many ways this was the opposite of the global network being developed by economists and politicians, which put a primary emphasis on trade and commerce. Their global network would force some areas to become permanent suppliers of raw or unprocessed materials and to put maximum emphasis on production for export to industrial or technological centers, which would then provide the manufactured materials for the original producers to purchase.

All these ideas contributed to my interest in *ecodevelopment,* meaning the improvements of living standards for people based on the use of knowledge derived from ecology to achieve sustainable ways of life, sustainable not only for people but for the ecosystems and species that had evolved on earth. I don't know who first used the term *ecodevelopment,* but it came to my attention in a speech by Maurice Strong when he accepted the leadership of UNEP. Many of us in IUCN agreed that it fit the programs we had been advocating. UNEP held a number of small workshops in Geneva to give better definition to the concept. In these I worked with Jimoh Omo Fadaka of Nigeria, Ignacy Sachs of France, Viktor Kollontai of the U.S.S.R., and others. The most complete coverage of the concept and its meaning came with a joint conference sponsored by UNEP and the UN Commission on Trade and Development at Cocoyoc, Mexico, in 1974. The Cocoyoc Declaration, written for the symposium by Barbara Ward (I believe), states the problems, goals, and objectives very well. The entire text is available in an appendix to my fifth edition of *Environmental Conservation* (1984).

Essentially, ecodevelopment stresses that human use of planet earth must respect the ecological constraints imposed by the natural environment. Failure to respect them will mean more and worse floods, more severe and frequent drought, more severe and frequent hurricanes and tornadoes on top of unpredictable levels of volcanic eruptions and earthquakes—essentially wiping out any economic gains. Since the world is an ocean planet, 70 percent seas and oceans, pollution and overfishing contribute to continuing disastrous effects.

Global warming caused by human use of hydrocarbons leads to all of the above disasters and yet we have been unwilling to take the necessary and known measures to cope with it. We appear, as a nation, to be in denial concerning this issue, or simply hoping that something will turn up to save us.

Second, ecodevelopment stresses that development must be directed to meeting the basic *needs* of the poorest people before paying attention to the *wants* of the elite. This is obvious but is the reverse of most economic development going on today. It has been noted that food flows from the poorest countries to the wealthiest who can afford to buy it.

Finally, ecodevelopment must emphasize development of self-reliance to break down dependency on the wealthier and more technologically advanced, taking into account local cultures and local environmental conditions. In my travels for IUCN and later, I have attempted to spread this message to many countries, from Australia and New Zealand to South Africa and Sri Lanka. It has not always been well received, since it goes against the interests of the transnational corporations and the exploiters. However, I keep trying.

While I was up to my ears in my favorite research areas, I was largely unaware of the growing antagonism of most of the staff toward the directorate. Already at the time of our general assembly at Banff, Canada, one of the problems surfaced. Frank Nichols, who was aggressive and well organized, gave the impression that he was running the meeting and that the director general was essentially his puppet. Both the president of IUCN, Donald Kuenen of the Netherlands, and I spoke to Budowski about this problem, and I presume Kuenen also spoke to Nichols. However, the problem persisted, as Budowski was away much of the time, visiting countries in which we had projects, which left Nichols in charge. This would have been acceptable had Nichols been less willing to undermine Gerardo's authority and more considerate of his staff.

By the time of IUCN's general assembly in Kinshasha, Zaire, the situation had totally deteriorated. On the flight to Zaire I was bombarded with complaints by other staff members about the behavior of our directors. I was told that Nichols had virtually terrorized some of the staff. Since all these people were dependent on their jobs at IUCN for, in most cases, support of their families, and were working in a foreign country, they were desperate. Furthermore, IUCN was reaching a penniless state.

The situation did not improve in Zaire. To begin with, it was very hot and humid. Our meeting place, constructed as a new and relatively luxurious locale for such gatherings, had an excellent auditorium and meeting rooms, but the food was terrible and drinks, with the exception of Zairian beer, were very expensive. Our quarters were not far from the Congo River, a vast moving waterway whose other shore was invisible, in another country. I thought of Kipling's "great grey-green, greasy Limpopo River, all set about with fever-trees." The river carried silt and pollution from millions of acres. I would not have dared to venture into the water. There was a haze in the air, no doubt smoke, which gave us spectacular sunsets, and at night a full red moon rising.

Unlike at any previous IUCN general assembly, we peace-loving conservationists found ourselves frequently encountering President Mobutu's special troops. Wearing new uniforms and armed with Uzis, Kalashnikovs, or whatever, they were omnipresent. Those who carelessly snapped their cameras at various sites around the meeting place had their cameras taken and film destroyed. On President Mobutu's first appearance at the general assembly, a throne covered with leopard skins was set up for his distinguished backside. An armed female contingent in the latest camouflage wear did a fancy Congolese dance and sang songs in his praise.

The motel rooms in which we were placed had been built recently by unskilled people. Showers, washbasins, and toilets did not necessarily work. Still we were able to adapt and went through the first technical sessions of the meeting without serious problems. Then, however, the VIPs were taken on a fieldtrip to the national parks, whereas the rest of us were left behind. Sitting in the bar in the evening drinking Congo beer and watching rats run around the wainscoting offered a perfect setting for rebellion.

I shared a room with Duncan Poore, former director of the British Nature Conservancy, who had joined me in IUCN as an additional senior ecologist. One evening we were invited to a meeting in another motel room with members of the American delegation to the general assembly. We were told that if the director general did not withdraw his nomination to continue for another three years, the United States would pull out of IUCN, and that it was up to us to bring about the withdrawal. If we were successful, we were told that a half-million U.S. dollars from an unstated source would be made available to support IUCN. Furthermore, we were given the impression that if the U.S. pulled out, the other major donor nations would soon follow suit.

This was all that was needed to blow off the lid and set the IUCN pot to boiling over. Emergency staff meetings went on into the night. Who advanced what idea I do not know, but it was finally agreed that a strike was called for. The formal general assembly meetings would be halted if all of us refused to do any further work, and we would refuse unless the director general agreed not to run for reelection.

I pointed out what happened to revolutionaries when the revolutions did not succeed, and that the proposed rebellion could succeed only if every one of us, from support staff secretaries to principal area chiefs, swore to stand fast, not give an inch, regardless of threats. I said that there could be no retreat. Furthermore, that each one of us, individually, would undoubtedly be taken aside, pressured and questioned in order to separate the leaders of the revolt. Fortunately, I pointed out, we had no leaders. Some time around then, I was asked to *be* the leader and to inform my old friend Budowski that he must resign. I protested and squirmed but finally had to agree that I was the only one close enough to Gerardo to have a chance of bringing a peaceful solution to the crisis. I was wrong.

While awaiting the return of the VIPs, I insisted that all of us circulate among the hundreds of delegates who had arrived to tell them what was happening and enlist their support. I took on the African delegates, in which I was aided by my old friend from Nigeria, Jimoh Omo Fadaka, who was well respected in Africa.

When the time arrived, I went to speak to Gerardo to give him the bad news. I tried my best diplomacy, but there is no gentle way to tell somebody that in his absence his staff had voted him out of office. He was furious and refused to believe me. I suppose he thought I was plotting to replace him in office. In any event, the struggle began. He talked to all his supporters in the general assembly and to each of the staff members. The head of the Zairian delegation offered to throw us all in jail. However, that not only would have stopped the general assembly but would have brought IUCN into the world news spotlight with publicity it had never been able to access before.

An emergency meeting of the executive board was called and we "ringleaders" were called before it. Nobody cracked. Nobody gave way. The U.S. representative, however, denied any threat that the U.S. would pull out of IUCN or that any such conversations had taken place. National delegates were spoken to about the problem. Finally, it was agreed that Budowski had to withdraw his candidacy for a third term. He did, as gracefully as possible. A special general assembly meeting was called for in Geneva, specifically to vote in a new director general and to change the ways in which directors general were to be elected in the future.

The night before we were all to depart for Brussels and Geneva, the staff held a celebration party. There was a great sigh of relief when the plane took off the next morning with nobody shot or jailed.

The whole incident was promptly swept under the rug as if it had never happened, and as if the director general's withdrawal of his candidacy were a normal thing. Frank Nichols, the deputy director general of IUCN, agreed to take early retirement and go home to Australia.

It would be naive to assume that all went well after the incident in Zaire. It did not. For one thing, we were still virtually penniless. For another, I was asked to serve as acting director general with the understanding that I was not interested in becoming a candidate for election at the emergency general assembly meeting. I agreed, with the understanding that Duncan Poore, who did intend to be a candidate, would assist me. He and I were put in the position where we had to beg funds from the World Wildlife Fund to meet the next payroll. It was humiliating. I do not know what wheels were turning at that time. The executive board was acting, however.

Meanwhile several of the mid-level staff decided that we were all overpaid, including themselves, and proceeded to plot to change the situation, a sort of counterrevolution within the revolution about which I was not kept informed. I do not know what motivated them. We were paid on the UN scale but our positions actually had more responsibilities than their UN equivalents.

I attempted to run IUCN in a democratic manner, seeking advice and opinions from all staff members. One European colleague accused me of having "silly American ideas about democracy." He assured me that Europeans had no such illusions. Harold Coolidge, whom I also consulted, since he was one of the original founders of IUCN, indicated that I was asking the "hired help" to run the house. Up to a point, I was willing to look the other way when my colleagues exercised their crusty old aristocratic attitudes and proclivities. When people's livelihood was threatened and their lives made miserable, I had to draw the line.

The final blow came one day when I was working in my office and Donald Kuenen, our president, and Luc Hoffman, the executive vice president of World Wildlife Fund, set themselves up in the office next to me and proceeded to call the staff in one by one. Under the implicit threat of possible dismissal, each one signed a paper accepting a 25 percent pay cut. I knew nothing of this until I was called in and informed that virtually all the staff had agreed to take the cut and that it was now my turn.

I told them what I thought of their undercutting my authority. I informed them that I had already taken a voluntary pay cut, as had most of the staff, that I would not sign their paper, that they could immediately have half my pay since I would go on a half-time appointment, and that as soon as I finished the projects I was working on, they could have all my pay, since I would be leaving. Furthermore, as of then I resigned as acting director general.

Nothing ever ends that dramatically, except perhaps life itself. I called a general staff meeting and informed them all of my action and my feelings. Then I went home to break the news to Elizabeth. Next thing our Australian board member came to our apartment and asked me to reconsider my words and actions. I told him, "No way."

So that is my story. I am sure that every participant in the "incident in Zaire" has a different account. I have felt badly about the whole thing and have wondered since if I had been led into a trap. If so, I can guess who baited it. I must say, however, that IUCN today, the new World Conservation Union, differs from what I have described. It has been changed in many ways and is no longer the lean, hungry, and confused organization that I once knew but rather well funded and no longer subject to the same pressures from World Wildlife Fund International. Much credit must be given to its recent two-term director general, Martin Holdgate of the U.K., but also to his predecessors.

My immediate replacement as acting director general was Duncan Poore. However, the emergency general assembly in Geneva elected the Canadian David Munro, who resigned after one term to be replaced by Lee Talbot of the United States, who resigned to be replaced by a fellow American, Kenton Miller, who turned the reins over to Holdgate and a new regime of relative peace and stability for the newly named World Conservation Union. At the 1988 general assembly in San José, Costa Rica, all former directors general and acting directors general met together without visible signs of stress to wish Holdgate the best of luck. All were elected "Members of Honour" of IUCN.

After completing his two terms as director general, Martin Holdgate agreed to write a history of IUCN. This he did, and it was published in 1999 by IUCN under the title *The Green Web.* He asked me to contribute to the chapter he called "The Night of the Long Knives," concerned with the turmoil leading to the replacement of the director general in Kinshasa. I sent him a draft of my chapter on "the incident in Kinshasa," some parts of which he included in his book.

19

# Return to the South Pacific

Oddly enough, some of my most interesting assignments for IUCN took place after I had left the IUCN secretariat but was still available for projects. One of these was a survey of the status of conservation in the islands of the South Pacific. It was essentially a follow-up on talks I had given, first in New Caledonia in 1971 at a symposium on the conservation of nature, then in Wellington, New Zealand, in 1975 at the first South Pacific Conference on National Parks and Reserves, and then in 1976, in Apia, Western Samoa, at the second such conference. The third of these conferences was to be held in Sydney, Australia, in 1979, where it was scheduled to coincide with the hundredth anniversary of the establishment of Australia's first national park, the Royal National Park located south of Sydney, and with a Southeast Asian Conference on Ecology and Development to be held in Kuala Lumpur in Malaysia.

Elizabeth and I took off on Singapore Airlines, going first to Singapore, where we would stay overnight and then catch a flight to Kuala Lumpur. There was time for us to see a bit of Singapore and to visit its "national park," which was essentially a city amusement park but did contain a small reserve protecting a fragment of the original Malaysian rain forest—not enough to maintain even a sample of the intact biota of a rain forest, but better than nothing.

The flight to Kuala Lumpur allowed us to see "the impenetrable jungle" that the island city of Singapore counted on to protect it from an overland attack during World War II. The Japanese went through it like a knife through butter and Singapore, with all its big guns aimed seaward, fell. Now in 1979 the rain forest had also fallen, replaced by oil palm and other plantations.

My talk to the conference was brief but stressed the need to have a firm base in ecology, the "eco-" in ecodevelopment. It is still worth repeating today when "sustainable development" is a stated goal for much that is ecologically unsustainable and going wrong.

We had the opportunity to renew friendship with Len Webb, the Australian rain forest authority who had, years earlier, taken me on a second visit to Lamington National Park, the scene for our 120-mile walk when I was with the 128th infantry regiment. In Malaysia we were able to go with him to one of the smaller national parks north of Kuala Lumpur and also to hear what was happening to the rain forests of Southeast Asia—definitely not good news.

We were supposed to leave Singapore on a night flight to Sydney, and at the airport we discovered a few errors our travel agent had made. First, we did not have visas for Australia; second, we were booked on a nonexistent flight to New

Caledonia after the conference in Sydney; and third, we were supposedly booked into a hotel in Noumea that had burned down several years before. Qantas Airlines came to the rescue and somehow managed to straighten out our itinerary. Fortunately, I did have an official invitation from the Australian government to attend and speak at the South Pacific conference.

Sydney was a second home for both of us, but we were shocked by the changes, since it had become a city of high-rises and freeways instead of the cozy city we remembered. Unfortunately, our schedule did not permit us to visit friends and relatives. We were taken on a special trip on a 100-year-old steam train to the Royal National Park, where a banquet was laid out for all the delegates to the conference. We did not see much of the park outside the picnic area where we all gathered but returned in the evening to our Sydney hotels.

At the conference the antagonism toward the French, then engaged in nuclear testing in the islands off Tahiti, was apparent. Though the official conference language was English, the French delegation insisted on speaking in French with an English translation. So the Papua New Guinea delegate in turn insisted on speaking in Pidgin with an English translation, and we heard such resounding statements as, "You no come bugger up forests belong us." It was a fascinating meeting.

At the reception following the first day of the conference there were some further events of interest. The Papua New Guinea minister of the environment had brought his wife and small boy with him. While I was talking to the minister, his little boy was trying to get his father's attention, since he wanted to be taken to the toilet. When the minister ignored the boy's frantic tugging on his pant leg, the boy finally, in desperation, bit him on the leg. But that was not the only incident of biting. The minister's wife had been taken on a shopping expedition to the big department stores in downtown Sydney, and showed up at the reception in a new outfit. She had left all the tags on the garments, which were fluttering in the breeze as she walked around. One of the Australian women who had accompanied her on the shopping trip stopped her to tell her that the tags should come off and offered to remove them. Lacking a knife or scissors, she proceeded to bite them off and handed them to the Papua New Guinea woman. But this was not what the Papua New Guinea woman wanted. The tags told the prices and types of material and she had been hoping to show them to her friends when she got home. Culture clash!

Our arrival in Noumea was not recognized by any of the government people whom I had hoped to interview. Though I had not been declared persona non grata there in 1971 (when the talk I gave prompted the newspaper headline, "Swiss go home"), the talk I had just given in Sydney had perhaps aggravated the earlier talk's effect and resulted in no welcome for me or IUCN in New Caledonia. Fortunately a friend of ours from California, Arthur Dahl, was now working in Noumea as the environment officer for the South Pacific Commission and was able to fill us in on developments. Yes, they were still strip-mining for nickel ore, leaving the land bare and the ocean polluted. There was no sign that the French company known as Le Nickel had any intention of stopping the mining regardless of its effects.

The Dahls invited us to dinner at their house, back in the hills above Noumea, an attractive house with a beautiful view.

After an enjoyable meal and conversation, Arthur drove us back to our beachfront motel and left us off there. He had already started for home when we discovered that we must have dropped our motel room key in his car. We went to the office, where the manager informed us that he had no other key to our room. He walked out with us to the front of the motel, where we could see our balcony, with a coconut palm leaning toward it. When all our questions about spare keys had negative answers, Elizabeth proceeded to shimmy up the palm tree, reached the balcony, leaped onto it, and entered the room through the balcony door. The manager was shocked. I am told he had new locks put on all the rooms soon after we left. Elizabeth was sixty-three in 1979. The world should be glad that she had not taken up a life of crime.

From Noumea we went to Fiji and stayed in a Holiday Inn in Suva, the capital. Our visit coincided with a severe tropical storm that brought extensive flooding and confined our travels mostly to the area around Suva. We did hear quite a bit about conditions on the islands from the biologist Birandra Singh with the National Trust. His biggest problem was that he had no boat and consequently could not keep up with what was happening on the over four hundred islands that made up the Republic of Fiji, except on those occasions when he could hitch a ride with the navy or with another governmental department.

Fiji suffered from some severe racial problems. The native Fijians, of Melanesian ancestry, were a majority population. Most of the other inhabitants were Indians, descended from those the British had brought over to work on their plantations, and from later immigrants. Most of the land was communally owned by native Fijians, leaving little room for the

Indians, who were largely confined to urban areas. The land issue seemed likely to become more of a problem as time went on. As we know today, it did.

On Western Samoa we were taken to visit the national parks on the main island and were impressed by what was being done. In particular, we visited one national park that ran from the mountaintops to the sea. The government had changed since my last visit, and I no longer knew the prime minister or the minister of the environment.

At our next stop, Tonga, we were to have an audience with the king, whose palace and grounds were centrally located in the capital city. We had been urged to do this by Harold Coolidge, who wanted me to pass on his respects and best wishes to His Royal Highness. The king was enormous, tall, and I would guess over 500 pounds, since he could not fit into a normal first-class seat on a jet plane. Since he was supposed to be on a diet, we were warned not to talk to him about food. We were to urge him not to expand the airport to enable the big 747 jets to land, since this would wipe out the only remaining natural forest on the main island.

The king was very open and friendly, but all he wanted to talk about was food. He maintained a big flock of geese on the palace grounds and extolled the virtue of goose egg omelets and other egg dishes. He did hear me out about the airport and the forest but was obviously convinced that Tonga needed the increase in tourism that the airport expansion would bring. All the women we saw in Tonga—totally brainwashed by the missionaries—wore ankle-length Mother Hubbards, with no skin exposed save their faces, hands, and feet. But Elizabeth lacked the prerequisite floor-length dress we were informed that she must wear for an audience with the king. Somehow she managed to buy the cloth and manufacture the skirt with her handy sewing kit the night before the occasion.

Tonga was the last stop on our tour. From there we went to American Samoa to catch the big jet back to the U.S.A. Because of my strong affinity for islands, and because they show in microcosm the issues our planet island must seek answers for on a major scale, I put together a report, "Conservation in the South Pacific Island Region." It went to IUCN and from there to UNEP in Nairobi, but I don't know of its having any effect on anyone. It may still be in somebody's in-basket.

20

# Back to the Land

There I was, sitting on a hillside where the ponderosa pine and black oak forests of the Sierra Nevada merge with the manzanita, ceanothus, and digger pine of the San Juan Ridge foothills. Once this area had been mostly ponderosa, but fires, timber cutting, and grazing had pushed the forest uphill and allowed the brush fields to advance. Now the pines were struggling to come back. But I was staring at a pile of 2 × 4 studs, concrete blocks, and miscellaneous building materials. I had in mind building a one-room shelter where I could stay while contemplating a more spacious and permanent structure. I was sitting in an area covered by kitkitdizze, or mountain misery, as some have called it. The temperature was well into the nineties, humidity was low, and there was no breeze. I did manage to force myself to lay out the concrete piers where I thought they should go to support the floor, but then I ran out of steam and couldn't push myself to do anything more.

"You shouldn't have tried to build anything on that hill,"

friends told me later. "There's something about its spiritual power that resists human interference."

Others remarked, "We tried to spend the night up there, but it felt too spooky. We had to move down."

The mythology surrounding that little hillside grew over time. Nothing got built there. I am inclined to go along with the facts: it was too hot, I didn't know beans about how to build a cabin, and the location was poorly chosen. But you can draw your own conclusions.

I'm not sure why I was in Nevada County at that time. I think I had a "home leave," a privilege of the United Nations' type of organization to which I belonged. Elizabeth had not come with me. If she had, there would have been a cabin sitting there before we left the scene. But I was alone with my pile of lumber. I did put together a decent campsite but no hut, house, or hovel.

All this was a result of ideas that came into my head during the turbulent period of the early 1970s. I was fed up with being a California expatriate at a time when all the social turmoil and possible hope for the future seemed to be centered in my home state. I was based in Switzerland, where nothing ever appeared to interfere with the orderly pursuit of money. I wanted to get away from the manicured Alpine scenery to the raw brush fields of home, where I might still encounter mountain lions and bears. When a person misses the feeling of chamise needles falling inside his shirt collar, he is really getting homesick. I found I was beginning to have nightmares about being caught forever in an endless UN conference in a city I did not know. I wanted a cabin way out in the woods.

In 1976 I went back to our land on the San Juan Ridge. We

had purchased 22 acres of a former cattle ranch that took over the site where miners looked for gold in the hills. The '49ers and cattle had gone, and the brush had advanced to cover most evidence of early occupancy. Initially we had no close neighbors, but soon thereafter Steve Beckwitt and family moved into the old ranch house over the hill, a half-mile away. Wildlife was everywhere, from fence lizards to occasional cougars and bears. Bobcats, coyotes, gray foxes, raccoons, skunks, and lots of deer lived there. Coyotes sang at night, and bobcat snarls mixed with owl hoots.

It was in many ways a magical place, with its oaks and pines, tall standing boulders of granite, dense brush fields, and open meadows. It was where the migratory wildlife of the Sierra Nevada came to winter and encounter the resident wildlife of the chaparral and oak woodland. I could see the Steller jays from up-mountain contesting territory with the resident scrub jays or see mountain quail and valley quail feeding on the same sources of plant seeds.

At times there I seemed to achieve a state of invisibility. I walked up to within 10 feet of a sleeping gray fox and told it to wake up. It went straight up from sleep to action and disappeared in a split second. Another time when I was sitting and meditating, a coyote started to amble up the trail toward me. He was evidently leading a family group, since I could hear others rustling through the brush. He kept coming until I feared he would bump into me before he saw me. I quietly said, "Hey, coyote" to him. He went up in the air and came down heading back down the trail. Then, when he was about 15 feet away from me, he stopped and turned to have a good look, perhaps to see if I was real. He then ran off with the others.

One evening I was sitting by the campfire with Sandra's big white malamute beside me. He stood up, started growling, and headed aggressively down the trail. Seconds later, he came charging back with his tail between his legs and hid behind me. Looking to see what monster had scared him, I saw a pair of small, glowing red eyes coming toward me until the firelight revealed a striped skunk. He detoured into the brush and went on his way.

Friends who knew what they were doing helped me put up a cabin in 1976. It lacked most modern conveniences but had a woodstove for heating, a propane cookstove, a propane-fired refrigerator, a sink with running water, and propane lamps. It had no bath or toilet—the former was the kitchen sink, the latter an outhouse. It had no electricity but did eventually have a telephone. It was not what I had intended, but it was OK. I could have lived there, but fate decreed otherwise. I joined the university faculty at Santa Cruz and became a part-time visitor to a cabin occupied full-time by one after the other of my three daughters, over a twelve-year period.

The experience of building and settling in a new area brought to me a realization of my inadequacies. I was supposed to be a pretty good ecologist, but I was a damn poor carpenter, plumber, electrician, road builder, and factotum. My technological ignorance was abysmal. My cultural level was basic hunter-gatherer, working toward hunter-gardener. More advanced levels of culture were mostly beyond me. I also realized my dependence on the continued functioning of our technological society. Sure, I could get along without electricity, but there was no way I could get along without a pickup truck, gasoline, propane, optometrists, dentists, and so on. In time perhaps, we could have grown most of our own food and decreased our dependence on the outside, but at what level would we then live? And who would be satisfied with that existence while knowing that the whole wide world, with all the rest of humanity, was sailing by to somewhere else, to a destination unknown?

Perhaps my daughter Lauren had the answer: "Why go into the woods to build a new community, disturbing the animals and plants that live there? You already have a community where you live. Why not make it function the way it should?" Or, to borrow an old saying, build a new society in the shell of the old.

Nevertheless I do not oppose the back-to-the-land movement. If you have been living in cities too long, it's an educational experience. It brings you back into contact with the realities of life and of the natural world on which we all depend. It can also make you face up to the problems of working with people who come from different home environments, different family cultures, and different ideas about how things should be done.

My colleague Dick Cooley, who was an old Alaska hand, tried to have public land set aside in Alaska where people could try living in the woods, either alone or with a nascent community of others. If they agreed to the basic rules of nature conservation, or in other words, if they agreed to live lightly on the land, it would give them an invaluable experience at little cost to the environment. Alaska still had space and wildlife adequate to support such social experiments. In fact, they were already happening in the back country, under no real control and illegally for the most part, on phony mining claims or in recreational pioneer life. Escape into the woods seems easier than rebuilding at home. But it isn't really. It just postpones the conflicts our planet is facing.

The abundance of wildlife we first encountered on our own land did diminish despite our efforts. We tried through controlled burning and limited brush clearing to restore the more open aspect the land had in earlier times. But the county was hiring professional trappers and poisoners to get rid of coyotes and bobcats, in the interest of sheep and chickens. Trappers did not come onto our land, but wildlife tends to range far. In later years we heard fewer coyote songs. Domestic animals played their part. House cats happily consumed white-footed mice and fence lizards. Even the best-behaved dogs tended to harass deer and keep smaller wildlife away. A 10-acre minimum lot size still allowed a lot of people to move into the area, one with a house 100 yards away. There came to be eight households within a half-mile of our place. This contributed to traffic and to wildlife decline.

Perhaps my greatest discomfort with what was happening in our area of San Juan Ridge was that it was part of a pattern of urban encroachment taking place up and down the Sierra. People fed up with city problems and having enough money to move away were beginning to crowd into Central Valley towns and cities where house prices were lower and living conditions better. Those who did not need to commute, and some who did, moved still farther out, to the Sierra foothills. Here, the greatest concentrations were building up below the 3,000-foot level, above which snow tended to accumulate in winter.

Unfortunately, the land they were, and are, building on is also the winter range for the migratory wildlife of the Sierra:

deer, cougars, mountain quail, coyotes, foxes, and Steller jays, for example. Winter ranges for deer are traditionally used areas to which the deer return year after year. Deer displaced by housing and disturbance move into less suitable habitat. Their numbers decline. Some deer and other species that find humans moving into *their* backyards behave more opportunistically. Newly planted gardens provide better forage for deer than old deerbrush, and cougars find it easier to pick off a poodle for dinner than to stalk a deer. Conflict occurs, and as usual, it is the natives that lose out to the newcomers. It's the American way.

In the real world, such as it is, human numbers, and the land and water used primarily for human-support purposes, continue to expand. Wild land and wildlife are pushed back into ever smaller areas. A process of insularization takes place as wild areas become encircled by tame. There have been many studies that show that when this occurs, wild species become extirpated and a dwindling of diversity takes place until only the most resilient species or those that require the least space survive.

If we mean what we say when we pass an act to protect endangered species, or a convention to maintain and restore biodiversity, then this steady encroachment on the land that is needed by wildlife must cease. Our species is already occupying too much space, and we must leave room for all the millions of other species that have found a home on earth. This is not just romantic idealism, but ecological realism. We depend on the continued functioning of the biosphere and its ecosystems, and we cannot replace those functions, except in small part, by human artifacts.

Yes, let us allow and even encourage new settlers, but let them settle on land already damaged by past misuse and have as their purpose the restoration of the wild, not the clearing of the wilderness.

In a more ideal world, wild land, in the broad sense of the word, would occupy the most space. Human settlements and the more intensive forms of land use would be restricted to smaller areas set within matrices of wild land. This is the pattern that characterized the villages and cities, fields and farms, during most of the period from the "agricultural revolution" until relatively recent times, a century and a half ago at most, in North America. This pattern can be sustainable over the centuries. Our present pattern is not. After all, humans are the planet's most dangerous pest species. They require control if the world is to be kept safe for all living creatures.

21

# Damming Paradise

Why is it that despite all the talk of international aid and technical assistance, the poor get poorer and the rich get richer? Why is it that throughout most of Asia, Africa, and Latin America resources are being ruthlessly exploited for immediate gain, so that forests are vanishing and woodland, scrub, savanna, and steppe are becoming barren wasteland? Millions of dollars funnel into international efforts at nature conservation but the gains are small relative to the ongoing losses.

These questions and others related to them led some of us to seek out approaches to development that fully take into account the ecological principles that will affect development, and the conservation imperatives that demand respect. Thus, in 1968 the Conservation Foundation sponsored a conference, with which I was heavily involved, on ecology and international development. We brought together experts from around the world who had first-hand experience with internationally assisted economic development projects, mostly in third-world countries. Usually they reported tales of woe, indicating full or partial failure of projects or ecologically disastrous side effects of major engineering schemes all because ecological considerations were not taken into account. The volume that presented the findings of the conference, *The Careless Technology* (1972), edited by Taghi Farvar of Iran and John Milton of the Conservation Foundation, really indicated the need for a separate book, a guide to the ecological principles that should be considered prior to implementation of any major development scheme. Shortly after I went to work with IUCN in 1970, I took on the task of working with John Milton of the Conservation Foundation to produce another book, to be entitled *Ecological Principles for Economic Development.* As a third author we added Peter Freeman, who had worked with U.S. AID on various projects. Also, since we were working with IUCN we had to take into account the comments of experts around the world associated with IUCN or its related agencies. You could say that the book, therefore, had around eighty authors and the final volume published in 1972 was not written in quite the way any one of us might have done if left alone. Nevertheless, it had more than a little success and was translated into many languages. Whether it resulted in any real accomplishments on the ground, in influencing the economists, engineers, and governmental decisionmakers, I will never know.

However, the opportunity to look into the extent to which the situation on the ground had changed presented itself in the mid-1980s, fifteen years after *The Careless Technology,* when I was invited to serve as a consultant to the National Park Service on a project in which they were involved in Sri Lanka.

The island of Ceylon lies to the south and east of the southern tip of India. It is almost connected by peninsulas and small islands known as Adam's Bridge. To some it is the original Garden of Eden where Adam left his footprint on top of the mountain known as Adam's Peak. It has had many names. Once it was Serendip, the happy, fortunate isle. Now it is known as Sri Lanka. It is comparable in size to the state of West Virginia or to Ireland, and like the latter island it is torn by a long civil war, marked by "terrorist" attacks and official reprisal. As in Ireland, the war is between the two dominant religious groups, the majority of Sinhalese, who are followers of Buddha, and the Hindu followers of Krishna. That both religions emphasize peace and nonviolence does not impede the bloodshed.

Sri Lanka is the home of an ancient civilization. When Rome was a simple village, the northern city of Anuradhapura was taking shape as the center of an island kingdom. To Anuradhapura in the third century B.C. came Mahinda, son of the Emperor Ashoka of India, bearing with him a sprout of the bo tree, the giant fig under which Buddha is said to have achieved enlightenment. The bo tree continues to thrive, as does the Buddhist religion. For 1,400 years, Anuradhapura thrived as capital of the northern kingdom. Then, under frequent attack by invaders from India, the capital was moved to Polonnaruwa, where it remained for another three centuries. Perhaps the most remarkable achievement of these early kingdoms was an elaborate system of dams, reservoirs, and irrigation canals constructed throughout the dry northern and eastern sections of the island. Some of the ancient reservoirs (tanks) are still in use today.

For reasons not entirely clear, the ancient cities of Sri Lanka became ruins, the irrigation works silted in, and dry tropical forest established itself to hide and overgrow the temples and palaces. When European invaders first arrived—the Portuguese in A.D. 1505—the great kingdoms of the past were remembered only by scholars. European rule was long lasting. Portugal controlled part of the island for 151 years, followed by the Dutch, who remained in charge for 140 years, and then the British, who succeeded in bringing the last independent kingdom, Kandy in the southern highlands, under their jurisdiction. They remained in control for 152 years. The British had the pleasure of rediscovering and beginning the restoration of the archaeological treasures of the ancient kingdoms, a task that continues today under the leadership of UNESCO, through its World Heritage program.

Tom Dale and Vernon Carter, writing in *Topsoil and Civilization,* attribute the collapse of the ancient irrigation system to one primary cause: cutting of timber from the forested highlands where the rivers that watered the dry plains of the north and east had their origins. "They cut the trees from the upland forests, causing runoff and erosion to be greatly accelerated. Because of neglect or barbarian invasion, floods and silt washed out or filled up the diversion canals, and with the reservoirs dry, famine depopulated the island. Civilization disappeared about A.D. 1200."

Following the collapse, forests reoccupied the land. Now, however, a new cycle of deforestation is well under way with only 7 percent remaining of the once luxuriant rain forests, and a major clearing of dry tropical forest taking place. Ironically, this coincides with an attempt to reconstruct and exceed the extent of the ancient irrigation works.

Today you may find two contrasting approaches to economic development. One is the standard Western model—a multibillion-dollar, internationally supported effort to reshape the land through construction of five major dams and an intricate network of canals. The other is a low-key, village-to-village approach based on tradition and religion, aimed at increasing self-reliance. The former is the Mahaweli Development Program aimed at capturing and controlling the island's largest river, the Mahaweli, and replacing dry tropical forest with rice paddies. The other is the Sarvodaya movement, started by a high school teacher, A. P. Ariyaratna, and his students in 1954 and in 1985 said to be reaching more than seven thousand villages.

It has been said that traditionally the villagers in Sri Lanka would build three tanks to accommodate their agriculture. The first, high up in the forest, would be for the wildlife, so that they need not visit the village fields. The second, also in the forest, would be for the use of the *chena* cultivators, those engaged in shifting (slash-and-burn) forest farming. The third would be for the village rice paddies. In each rice paddy, a small section at the end of each field would be reserved for the use of birds. When asked how the birds knew which areas were theirs, a traditional farmer answered: "We have been doing this for thousands of years. The birds have had ample time to learn which was their paddy and which was ours; they rarely trespassed onto our part of the paddy fields unless of course they were invited to do so to eat the paddy bug of the godewella worm—and besides, they would be chased away by the children."

The idea of reconstructing major irrigation works in Sri

Lanka has been a dream since British colonial days, and with independence in 1948 it became a goal of various Sri Lankan governments. In 1968 a joint study by FAO/UNDP came up with a proposal calling for fifteen dams and reservoirs along the Mahaweli to be built over a period of thirty years. The structures would store 6 million acre feet of water, and their installed electrical power output would amount to 500 megawatts. However, early development was slow, and in 1977, when the government of J. R. Jayawardene came into office, a new accelerated program was initiated. Five dams were to be built in six years. One hundred seventy-five thousand people would be employed in construction, three hundred thousand would be settled directly on the newly irrigated farms. One hundred fifty thousand would be employed in agricultural service jobs—in all 1 million people or 6 percent of the total population would be moved into the project area. Initially some 180,000 hectares were to be irrigated, leading eventually to 900,000 hectares of irrigated land. This was to be divided democratically with first priority going to those displaced by the project, but with each family receiving only 1.0 hectare (2.5 acres) of paddy land and .2 hectare (1/2 acre) of homestead land (where houses and gardens were to be located).

Homesteads were to be clustered in hamlets of five hundred to seven hundred people, each with a school and dispensary. Each four or five hamlets would have a village center with a post office, high school, and markets. Land obtained under the project could not be sold during the first five years, after which it would become totally private and available for sale.

Though there were many warnings about the problems likely to be encountered, the accelerated Mahaweli program has gone forward. The first of five dams, the Polgalla diversion, had been completed before the accelerated program began. The others, despite unexpected difficulties and delays, have been constructed. The first to be finished was the lowest in the system, the Madura Oya rock-filled dam, built by the Canadians and completed in 1982. Ironically, it stands above an original earth-and-rock-filled dam built by the early kingdom at least 1,500 years before. In 1984 the British finished the 400-foot Victoria Dam near the city of Kandy, which flooded 9,000 acres of rich bottom land. In 1985 the Swedish construction company finished the dam closest to the watershed, the Kotmale Dam, 285 feet high. In 1986 West Germans completed the Randenigla Dam. To clear land for the agricultural enterprise, one-fifth of Sri Lanka's tree cover will be removed. Since this is habitat for Sri Lanka's remarkably diversified plant and animal life, the ecological impacts are already enormous. Led by the elephants, Sri Lanka's wildlife has "struck back," to damage the new rice fields and at times to endanger the settlers. But in this kind of contest, wild animals don't win.

It is inevitable that in massive irrigation schemes of this sort, success depends on centralized authority. Without central control, water cannot be provided when and where needed, nor can its distribution to all who need it be assured. Without central control the massive irrigation works cannot be started, let alone completed. Hundreds of thousands of people cannot be moved and resettled without central government's authority providing the controls and the means. Thus power tends to reside in the Mahaweli Authority, the ministry of Mahaweli development, which answers only to the president.

Inevitably, without a stronger central control than is yet apparent, ownership of the farming land will fall into fewer hands. Those with financial reserves will be able to weather the bad years and buy out those who lack those reserves. The family-clan networks that enable small farmers to survive in their old villages will not be present in the new hamlets and villages of Mahaweli.

It is difficult to envision an approach to development more different from the Mahaweli program than Sarvodaya. It had its beginnings in an exclusive Buddhist high school in Colombo, where Dr. Ariyaratna decided that his students needed more exposure to the life of the country's poor peasants. With permission of the villagers, he brought his students to work with them as a vacation project. The students lived with the villagers, shared their food, and worked with them planting gardens and digging wells. Villagers and students alike benefited from the experience. The idea spread to other villages and other schools and became a nationwide movement. Meanwhile, Ariyaratna traveled to India to learn from the followers of Mahatma Gandhi, particularly Vinoba Bhave. He brought back to Sri Lanka Gandhian methods and the name Sarvodaya Shramadama for the movement, taken from the Gandhi vision of a society based on self-reliance, truth, and nonviolence. The emphasis shifted from education to rural development, from one primarily involving the schools, to one organizing the villagers to help themselves.

The aim of the movement is for each village to satisfy the ten basic needs of its people: water, food, housing, clothing, health care, communication, fuel, education, an environment that is clean, safe, and beautiful, and finally a spiritual and cultural life. In each village the movement starts with a village meeting organized usually with the resident Buddhist monk and held most often in a temple or religious center—this to encourage participation by the women, who do most of the work. The Sarvodaya organizer explains the purpose of his visit and challenges the villagers to identify an important priority for the village and join together to achieve it. At first this is usually something relatively simple such as building an access road, clearing a tank or well, planting a garden. The movement supplies ideas, skills, and, when needed, credit and materials. Gains are seldom spectacular, progress is often slow, but the practice of working together for the common good brings feelings of achievement and recognizable material gains. The value to the villagers is evidenced by the popularity of the movement, reaching seven thousand villages in Sri Lanka by the mid-1980s and spreading overseas to Africa and other areas. Obviously, a movement of this kind needs support, and this has been available from religious and charitable organizations in Europe, but on a small scale and not at all comparable to even one of the Mahaweli developments.

What does Sarvodaya think of Mahaweli? Ariyaratna's answer is:

> We are opposed to large-scale projects of this nature. But unfortunately this project cannot be reversed now because it is already being built. So what we have to do is to eradicate or minimize the evil effects of this project. Large-scale projects of this nature bring about large-scale corruption, and nobody can deny that. Large projects disturb nature in a big way, without knowing what the consequences will be. It may be easy to do large-scale irrigation, but while water is the life blood it can also be the

> cause for elimination of human civilization if not controlled properly. If Sarvodaya had any hand in it we would rather have renovated all the old tanks and wells and other irrigation systems in this whole country at a lesser cost. We would have liked to see small projects rather than huge schemes.

Obviously the relationships of Sarvodaya with the Sri Lankan government have been mixed. Some say Sarvodaya itself is now too big, that it has lost touch with its grass roots. However, in 1987 Sarvodaya received a grant of $17 million from the AID. It remains to be seen if Sarvodaya can survive this kind of assistance.

It would seem that the official, government-sponsored Mahaweli project was simply setting out to repeat all the serious and damaging side effects of the projects we had reviewed in 1968. However, there are areas of difference, and that was why I was there as part of a National Park Service team. To help compensate for or mitigate the effects of the Mahaweli project on the dry forest environment, the Sri Lankan government proposed to establish a number of new national parks and to enlarge some of the existing parks. We were called in to advise and assist with this aspect and to train the new national park staff.

During our initial visit to Sri Lanka we first looked over the national parks that existed and the areas proposed for new parks. We spent some time in Wilpattu National Park on the northwestern coast, a long-established reserve noted for its wildlife diversity. Unfortunately, we discovered among the national park personnel what we called a "fortress mentality." The park boundaries were to be strictly patrolled and all people were to be kept out except those who came by car to the main gate. These, too, were carefully controlled, not allowed to drive around without a game guard, not allowed to leave their cars except in a few designated areas. Admittedly, there was dangerous wildlife—elephants, leopards, sloth bears, wild hogs—but the restrictions seemed to focus more on controlling anyone who would damage the park and its wildlife than on protecting visitors. There was virtually no effort to encourage the level of tourism that would build a popular base of support among the people of Sri Lanka.

We spent some time with the park staff at Wilpattu talking about new approaches to park management. However, a short time after we left Sri Lanka most of those we talked to were dead. A group of Tamil Tigers, the guerrilla arm of the Tamil liberation movement, swept through the area. They carried out a massacre of pilgrims and staff at the Buddhist shrine in Anuradhapura and then, on their way to the coast where they would be picked up by boats, they passed through Wilpattu. Just in passing they gunned down more than twenty of the park staff.

By the time of my next visit, in August and September 1985, the situation had deteriorated. The northern part of the island was considered unsafe. I carried out my seminars and field trips in Yala National Park at the southern end of the island. The only trouble we encountered was from a leopard who had bitten one of the guards and a local fisherman shortly before our arrival.

The civil war grew worse after my departure. The Indian army was called in to assist the Sri Lankan defense forces, but nothing was solved. Violence continues today. To say that civil war is destructive to nature conservation is to state the obvious. Positive accomplishments of any kind tend to be held in abeyance.

The Sri Lankan experience made me once again aware of the problems involved in any sort of intergovernmental work. I was there as part of the park service team representing our government and assigned to work with the Sri Lankan government's Department of Wildlife Conservation. This Sri Lankan department was in charge of the national parks. However, as we drove around the island we encountered signs designating biosphere reserves, part of the UNESCO system of protected areas. Our Department of Wildlife Conservation guides either knew little about these reserves or were not willing to talk about them. They were, as we found out, under control of the Forestry Department and apparently not to be considered in our work. One of these was the only protected rain-forest area in the country, but it was out of bounds for us.

Furthermore, we discovered in passing that some of the new rice fields developed with the Mahaweli Development Program were on the immediate boundary of the proposed new national park. Trouble was built into this location. Elephant damage to the rice fields and people, along with less spectacular damage from other wildlife, was inevitable.

Another irony we encountered was that I could not get any information about the activities of Sarvodaya from any of the government people I talked to. I tried hard in private conversations to discuss this topic with some I considered good friends, but I got nowhere. Apparently at that time cooperation with Sarvodaya was not to be approved by this department of the government. Obviously, some other governmental department had a different view or the U.S. AID grant to Sarvodaya would not have been approved. And since my time in Sri Lanka belonged to specific governmental groups, I did not meet any of the Sarvodaya people.

Some of my final remarks, which attempted to set forth ideas on the role of protected areas in rural and national development, were given at a seminar held in Colombo (September 1985) and are repeated here. They may indicate why I was not invited back. Or perhaps it was the civil war.

> We know that in the long run a nature reserve will not be effective in conserving nature in a hostile environment. A nature reserve cannot be maintained in a healthy, functioning state if it is surrounded by impoverished or hostile people. A nature reserve cannot be maintained if it is surrounded by a degraded and deteriorating environment. You cannot maintain it in a functioning order if land use practices in the surrounding environment are inimical to survival of the species within the reserve.
>
> Planning, control, and management of land-use in the area surrounding the reserve is almost as important to the survival of species within the reserve as protection and management of the reserve itself. This means that buffer zones, corridors, timber or fuelwood plantations, and wild land patches must be established or maintained, and their use for direct commodity production be kept compatible with the overall purposes of nature conservation.
>
> Economic development must take place in the communities surrounding the reserve, and this development must be tied, in part, to the existence of the reserve. Reserves must be seen as places for employment and potential sources of income by the people living near or directly

affected by the reserve. The reserve must be seen in a favorable light. Visitation of the reserve, particularly by schoolchildren, must be actively encouraged. Reserve staff must be looked upon as friends of the people, not as hostile forces.

Obviously, the Department of Wildlife Conservation cannot by itself accomplish these two major tasks. Other governmental departments, local authorities, nongovernmental organizations must be involved. But the department must take an active role, since nature conservation is its primary mandate. Nature conservation, however, must also be a responsibility for all planning agencies and all agencies concerned with land and resource development. Accomplishing this level of cooperation and coordination may be the most important single goal for the future of wildlife and the wild heritage of Sri Lanka.

22

# Other Ways of Life

Sometimes people are not aware of what is going on around them. I fear that has been a near permanent condition with me, and I can give you many examples. However, my Sri Lankan visits brought one of these experiences to my particular attention. In traveling around the old and proposed national parks, I did hear some mention of the Vedda, but their significance was not apparent. A year later, while on the board of directors of the World Wildlife Fund, I received a copy of a letter from Hayden Burgess of the World Council of Indigenous Peoples, asking our help in saving the remaining indigenous people, the Vedda, and their land and culture. Their lands were about to be included in the proposed Madura Oya National Park, and according to the rules set up by the Mahaweli Authority, the Vedda must be moved out.

The Sri Lankan flag features a rampant lion but there are no lions in Sri Lanka. There are, however, or there were lions in northern India, and it is from there that the Lion people came, the present Sinhalese majority on the island. Long before these invaders from India arrived, the Vedda had lived for many thousands of years in Sri Lanka. They are an ethnically distinct folk with features in common with the Australian Aborigines. They were hunters and gatherers who thrived in the biodiversity and tropical richness of the island of Ceylon. They were no match for the invaders' military and retreated to the eastern jungles, where they continued their ways of life. Until the Mahaweli project came along, nobody bothered them very much. They had their own villages and lived apart from the agricultural lands.

It should be noted that the ecological variety and species richness that have made areas of Sri Lanka fit to be considered as national parks have survived over millennia in the presence of the Vedda, who have lived as part of the natural environment. They have never threatened the existence of the elephant, the leopard, or the bear. Yet the rule that says that people may not live in a national park (except for those employed by the park authority) has caused the Vedda to be moved out of their traditional villages and resettled in one of the Mahaweli new villages, well removed from their forest home. Cultural breakdown accompanied by disease, alcoholism, and drug dependence is almost inevitable. Yet in the government's view, these people are to become paddy rice cultivators—a role they have carefully avoided for centuries. The simple solution of employing the Vedda as park caretakers apparently had not even been considered.

With the loss of the Vedda way of life, the equivalent of a full library of knowledge, passed on by oral tradition from one generation to another, will now be lost. Only they know the animals and plants of their forest with a detailed intimate knowledge of their values and uses. I have been amazed personally by a tracker employed by the Sri Lankan Department of Wildlife Conservation who shared to some extent the Vedda's skills. He lacked formal education, but he could follow tracks where technically trained experts could not even see them. He knew what the animals were doing, and why. He had names for all the plants and all the animal species we encountered, and stories to tell about them.

Once in New Guinea in a native village, taken over as headquarters of the 32d Infantry Division, I was sitting by the door of my hut looking out on the central compound around which the native houses had been located. Out of the surrounding jungle a huge black wild boar arrived and walked into the center of the village. He did look mean with his long white tusks showing. He created a chaotic response. All of the soldiers fled, many climbing ladders into the huts. These were heavily armed men with all the latest weapons, but one wild boar made them flee. From somewhere, a New Guinea man arrived, armed only with a fire-hardened wooden spear. He ran at the boar and threw his spear. It went right into the boar and killed it on the spot. Some of his friends then came and they carried the boar off to provide an evening feast at their camp. We warriors of America were left to brood about our reaction to an encounter with nature. No doubt the New Guinea folk talked about it too.

During the 1940s there were still many people who had no contact with Western civilization. Speaking hundreds of different languages, each tribe led its separate life in the mountains or along the coastal rivers. Now the old ways have mostly broken down. While we were in Buna exchanging bombs, artillery rounds, and small-arms fire with the Japanese, an Australian patrol was ambushed by headhunters in the Owen Stanley mountains. Most of them were killed by spears or arrows. An unusual way to die in a world war. The nation of Papua New Guinea has been created from what was once the Australian Mandated Territory, whereas Indonesia has taken over the western half of the island and christened it Irian Jaya. In Irian Jaya the original people have been displaced to a considerable degree and forced to give up their old ways. Immigrants have been brought in from Java to replace them on the land. It is the same old story of conquest and colonization with displacement or elimination of the indigenous people that has taken place all over the world and continues to occur.

I first really became aware of the conflict between conservationists and the rights of indigenous people when I first went to work for IUCN. I had been asked to take a critical look at the international standards criteria for areas designated as national parks, since there was a great deal of confusion about the use of the term. In America national parks had mostly followed the Yellowstone model—large reserved areas owned and protected by the highest national authority, in which visitors were to be welcomed provided they did no damage to the natural scene. Permanent residents were ruled out, and those who already owned land within the established boundaries (inholding) were to be bought out and moved as financing permitted. Yet in Great Britain national parks were of a different nature and could include farms and villages. They were essentially managed landscapes, incorporating rural farmscapes as well as natural areas. Those areas more equivalent to American national parks were known as nature reserves and did not necessarily encourage visitors.

IUCN had been charged by the United Nations to set the international standards for parks and reserves. The *United Nations List of National Parks* (and equivalent reserves) enumerated these criteria and those areas meeting the criteria. However, around that time I read Colin Turnbull's book *The Mountain People* (1972), in which the cultural breakdown of the Ik tribe of Uganda is described. When the Ugandan government decided to establish the Kidepo Valley National Park, it ruthlessly displaced the Ik people, who had always lived in the Kidepo. The riches of the Kidepo's flora and fauna were certainly preserved in—and perhaps by—their presence. They were totally unfitted to survive as part of the pervading Ugandan culture into which they were forced to move. Reading further, I encountered many similar stories, such as the displacement of the Waliangulu, the elephant people of Kenya, to make way for the Tsavo National Park. Of course I did not have to leave my home country to find examples—the story of the Yosemite Indians, though their displacement occurred before the park was created, fitted the pattern. Those who had long occupied the land and whose continued presence permitted wildlife and the wild scene to thrive were displaced almost everywhere.

Suffice it to say the UN criteria underwent changes over time, but the changes that permitted the continued occupancy of indigenous people in national parks were either not read or ignored for other reasons in Sri Lanka as well as other countries. I have noticed no tendencies to welcome the native people back into the national parks of the "lower 48" United States, though indigenous rights are being respected in Alaska.

Still, the issue is not so easy to resolve. What happens when indigenous people pick up some of the more despicable habits of the prevailing world technoculture? When they swap their spears and arrows for Kalashnikovs and Uzis, are they still to be regarded as hunter-gatherers suited to life in a nature reserve? The Kayapo Indians of the Amazon, subject of so much research by anthropologists and idolized by conservationists, have decided to sell their mahogany timber to the highest bidder in order to buy artifacts from Rio and São Paulo. Is that OK? And among the many Native American peoples who still have land they can call their own, there are those who want to sell their fossil fuel reserves or any other resources with cash value in order to gain more dollar income and permit "modernization" among their people. The fact that they are opposed by the traditionalists only mirrors the same struggle taking place in the prevailing culture between conservationists and developers.

The answer for nature reserves in general and particularly for biosphere reserves seems to be that emphasis should fall not on a particular ethnic group and culture but on a way of life, whoever practices it. Certain ways of living have proved to be not harmful to the aim of protecting the biodiversity of an area. A population limited in number, practicing a way of life shown to be not harmful to the conservation of nature can be permitted to live in protected natural areas. If numbers increase or the tendency to use resources more intensively develops, the excess numbers or those who wish greater exploitative uses should move out to other more suitable areas. This, of course, is much easier to say than to carry out.

But it would do no harm to apply the same rules to everyone, everywhere, and not only to nature reserves. The earth is the only known nature reserve in the entire universe. It is the only place where the diversity of life as we know it can survive. Should we not limit numbers of people and levels of land use and resource exploitation to what we know can be sustained without harm to the biodiversity of the planet? I do not know where to send those who do not conform. Outer space, perhaps?

## 23

# Biosphere Reserves

When I retired from the university in 1989 and no longer had to meet a regular schedule of classes and faculty activities, I was able to begin some projects more directly related to my interests. I became more active in the Earth Island Institute and was asked to participate in a joint expedition with Russian scientists to Lake Baikal in Siberia. At that time Mikhail Gorbachev was president of the Soviet Union and was instituting his new policy of openness to the outside world. My friend David Brower, chairman of Earth Island Institute, was going along with his wife, Anne. Fran Macy, who was head of a U.S.A.-U.S.S.R. group, was arranging our meetings with Soviet scientists and working out our travel plans. I was interested in seeing Lake Baikal and also the wilderness surrounding most of the lake and hoped to see some, at least, of the Siberian wildlife.

Lake Baikal has been called the Sacred Sea of Siberia. It is of great spiritual importance to many of the Russian and Siberian people as well as of great biological interest. It is said to be the world's largest freshwater lake—noted for the purity and clarity of its water and for the great number of species found nowhere else, including a freshwater seal—and is more than a mile deep, containing more water than all of the Great Lakes of North America. Now, however, Lake Baikal has become of great concern internationally because of increasing levels of pollution entering the lake as well as the continued hunting of the seals and heavy fishing pressure on some of the lake's fish species.

We were to travel by hydrofoil from the southern area of the lake, near the city of Irkutsk, to the northern end of the lake at the town of Severobaikalsk. Somewhere outside my hotel window in Severobaikalsk a great number of dogs, presumably sled dogs, were kept. I was kept awake by the racket they were making, yapping and howling. Then from somewhere in the forest north of the lake came the long howl of a single wolf. The dogs were instantly quiet. I felt that I had been welcomed, even though I was not to see or hear another wolf during our visit.

The Soviet Union appeared to be in competition with the United States in the establishment of biosphere reserves. As I recall, each submitted twenty proposed areas for biosphere reserve status in 1974. In both cases the reserves were to be formed from already existing protected areas, mostly strict nature reserves for the U.S.S.R., and mostly national parks and Forest Service research areas for the U.S.A. American advisers also came up with the idea of cluster reserves, made up of separate units that might have different purposes but together would meet the recommended status of biosphere reserve. One of the larger Soviet reserves was on the northern and eastern sides of Lake Baikal. It was an impressive area on the map, but on the ground it seemed to have little protection, since the person in charge had neither the necessary transportation nor the staff to patrol the area, which was mostly roadless.

In the United States at that time we were encouraged to think of the U.S.S.R. as a police state. I was therefore impressed to see how few police were around. I saw none in Moscow, though I did see a number of soldiers near the Kremlin. They seemed to be mostly engaged in selling parts of their Red Army uniforms and insignia to tourists. In the wilds of Siberia there seemed to be none of the usual protective laws or law enforcers. People camped wherever they pleased, built fires anywhere, cut down trees and killed animals whenever they felt inclined. Siberia was more like America's wild west than the United States today.

The high spot, in more than one sense, of my visit to the Baikal region was a helicopter flight over the wild areas north and east of the lake. I was determined to see any forms of wildlife that were present in this forested land, so I kept the porthole open and my head exposed to the full blast of air, in order to see as much as I could. This act had consequences that I did not expect. I saw impressive mountain scenery and a lot of Siberian taiga forest. We flew low enough for me to have seen anything from a wolf to a moose, but I saw no large mammals. They could have been hidden under trees, but most likely they were not there. We had an excellent Russian pilot, but at times I thought we would scrape the rotors on the cliffs.

My Russian interpreter seemed to be suffering from bronchitis and I doled out aspirin to help him get through the day. It was not long after my helicopter journey that I developed the mother of all nosebleeds. It was a major hemorrhage, not to be stopped by the home remedies I and my colleagues tried. Finally somebody called for an ambulance. It arrived quickly with a Russian woman doctor on board. By then I was in and out of consciousness, mostly out. I remember her injecting me and sending a large amount of some fluid into my veins. I was taken to a hospital where a male Russian doctor packed my sinuses with some stuff. The hospital staff who took care of me were all Buryat Mongol women, and though I do not know what they said or did, I woke up the next morning, still alive. The doctor, after removing whatever he had put into my sinuses and seeing that the bleeding had stopped, took me back to my hotel in his own car. I certainly could not complain about the level of medical attention I received in the Autonomous Buryat Soviet Socialist Republic of the U.S.S.R.

However, thanks to my interpreter, I started to come down with bronchitis that soon developed into pneumonia. I arrived home safely after having flown most of the way around the world and was taken to the hospital almost immediately for a week-long sojourn. In retrospect, I wonder if the wolf who called when I first arrived at Baikal was not welcoming me but saying "Yankee go home!"

Though I contributed little to the plans and recommendations that resulted from our visit, we did agree to establish a joint Russian-American program called Baikal Watch to assist in the protection of the lake. We recommended that the entire region of Lake Baikal and its watershed be designated as a biosphere reserve under the UNESCO program. Some of the members of our group were to return to Baikal to assist with the conservation activities.

Of the many programs included in UNESCO's MAB program, the most successful, at least on paper, has been the biosphere reserve project, which led to the establishment of more than three hundred biosphere reserves internationally, with forty-seven located in the United States. These reserves are protected areas intended to maintain at least representative portions of each of the major natural communities of the earth. Each such reserve should include a carefully protected core area to maintain a relatively undisturbed ecosystem. And such a core would be surrounded by buffer zones, which could be used in various sustainable ways, with the intensity of such uses grading from the least intensive, such as ecotourism, to the more intensive, such as sustainable forestry farther away. Such buffer zones would protect the core from incompatible uses in its immediate vicinity. Thus clear-cutting of forest would not go on next to an ancient forest core, and a protected core of tall grass prairie would not stand next to a severely overgrazed grassland.

In establishing a biosphere reserve, its staff should take care to involve the local resident population, along with regular or traditional users of the reserve, from the start of planning so that there would be no uprooting of people from their homes or exclusion of long-established users of the area. Thus, the biosphere reserve could hope to draw on local support and user support alike, enlisting many de facto protectors who would discourage or prevent incompatible activities. Employment of local people as rangers, wardens or assistants in managing the reserve would also be appropriate in many countries.

It also had to be made clear from the start that biosphere reserves were in no way a takeover by the United Nations of territory belonging to the sovereign states. UNESCO would have no legal authority but would serve as an adviser or coordinator for the program, helping national or local governments in the establishment or management of the areas. The program could also provide models and standards that were accepted internationally for biosphere reserves and could, in theory at least, decide whether or not a particular reserve would be part of the biosphere reserve network. UNESCO could, if funds were available, provide expert advice on particular management problems, for example, the suitability of prescribed burning for maintaining open pine forests, the use of predator control to protect an endangered species, or the selection of species to be planted to help restore wetlands.

I became a supporter and advocate for biosphere reserves before the first ones were established and for a long time followed the progress of this endeavor. After I retired from the university and had returned from Lake Baikal, I was asked to take a more active role. This was with the nearest biosphere reserve to my home in Santa Cruz.

The Golden Gate Biosphere Reserve was established in 1988 to aid in protecting the natural areas of the central coast of California. It consisted of thirteen units protected by nine different agencies including federal, state, county, university and private agencies or organizations. By far its greatest area is in the ocean, managed by the National Marine Sanctuary program of the National Oceanic and Atmospheric Agency. On dry land the National Park Service controls the greatest area. Essentially the biosphere reserve extended from Bodega

Bay in the north to Point Año Nuevo in the south, and from the tops of the coastal mountains to the depths of the Pacific Ocean, including the Farallon Islands, a spectacular breeding area for seabirds.

In 1990 it was decided that the reserve needed an association, nongovernmental and nonprofit, to assist in promoting cooperation and coordination among the separate units of the reserve.

In developing countries there was often strong opposition to setting aside territory for national parks, since these often excluded the people who had traditionally occupied the park areas. However, biosphere reserves, if established with local cooperation, did not exclude human uses but established both the kind and level of uses agreed upon. Also, the prospect of coordinated scientific research was an important purpose of the bioregional reserve system worldwide. This caused the biosphere reserves, at least in some countries, to be placed in a different, more science-oriented department of the government from the national parks and other conservation-oriented agencies.

The Golden Gate Biosphere Reserve Association (GGBRA) was created in 1990, and I became a member of its first board of trustees. The president and one of the founders was Laurie Wayburn, who was also in charge of the Point Reyes Bird Observatory, the principal bird research organization on the central California coast. It was agreed by all that there should be a full-time salaried executive director and secretary. But finding the money to pay for staff was not easy, and for the first year we struggled along with a half-time person to fill both roles, and after that we had none. This meant that unpaid board members, most of whom were employed full-time elsewhere, had to fill these roles.

Laurie Wayburn resigned as president and later as board member when she left Point Reyes to take over a new organization, the Pacific Forest Trust, based in Boonville. She was succeeded by Sally Fairfax, a professor in natural resources at UC Berkeley. She carried out one of the first successful projects attempted by the GGBRA, a conference that helped define the role of the association in environmental education. However, there was no money available to carry out the recommendations or to pay for a suitable person to work on the projects.

Fairfax was unable to continue with the board because of increased administrative duties in the university. When she resigned, Marc Hoshovsky of the California Department of Fish and Game took over. He got off to a good start with a meeting in which the managers of the separate units came forth with a prioritized list of what they believed the Golden Gate reserve should accomplish. Unfortunately, he was forced to resign because of the pressures of his work with Fish and Game. During both the Fairfax and Hoshovsky presidencies I found myself in the role of acting president. So I finally agreed to accept the board's wish and take over as elected president.

Thanks to vice president Nona Chiariello, who did most of the work, we were able to carry out the managers' wishes and convene a major conference on biodiversity and begin to define just what species of plants and animals were or should be found on the lands or in the ocean and streams of the reserves.

Representatives of sixty organizations attended the symposium in December 1995, and participants generally agreed that it was a success. And it was also a strain on everybody's time and energy and required us to raise many thousands of dollars for the conference and its publication. When it was over, we were essentially broke as usual.

At that point I realized I had been an on-again, off-again president for some seven years: it was time for somebody else to take over the leadership. Philippe Cohen from Stanford's Jasper Ridge Biological Preserve agreed to carry on as acting president. All the GGBRA's acting leaders agreed that we must, if we were to continue, find the money to hire a full-time salaried executive director and secretary and provide them with an office and equipment along with operating funds. As of now, there has been some success and sufficient money has been raised to hire an executive coordinator.

It has been obvious since the idea of biosphere reserves first surfaced that they cannot succeed without funding. Governments may designate biosphere reserves, but without funding, the reserves are only signs by the roadside or plaques hanging on the wall. Nothing new will happen. And it is unfortunate, but true, that the agencies that could form integral parts of a biosphere reserve are usually themselves strapped for funds sufficient to take care of the state park, national park, marine sanctuary, or other reserve. They might wish to provide funds for the biosphere reserve but are unlikely to put money they need to hire another ranger—or, less glamorously, fix potholes or provide new latrines—into something that, while it might be more important in the long run, serves no immediate need.

Add to the above the apparent bitter hatred some in Congress feel about the United Nations and its agencies. Each year some congressperson sponsors a bill to abolish biosphere reserves simply because they started in a UNESCO program and can therefore be presented to the voting public as an attempt by the UN to take over the country. Since many, if not most, voters have never heard of the biosphere or such reserves, they may believe they are rallying to the defense of the U.S.A. against foreign invaders.

24

# Finale

I suppose it all started in 1989. It was a bad year in many respects, the year of the big earthquake when downtown Santa Cruz was knocked apart. The earthquake didn't bother our house at all, though it scared the wits out of us. More scary was the news I had a large aneurism on my abdominal aorta and could drop dead any time. But I didn't, and after a major operation I was more or less as good as ever. I had retired from UCSC that year, when I reached the age of seventy. Had I gone on teaching I would probably have dropped dead before long.

The most important event of 1990 seemed relatively insignificant at the time. Elizabeth was bitten by a tick during one of our many expeditions to Butano State Park for sketches and photos of redwood forest scenes. The tick in question proved to be the Pacific black-backed tick, carrier of Lyme disease, and soon after the tick was removed she developed the classic symptoms of Lyme disease. However, at that time it was apparently the policy of the Santa Cruz medical establishment that there was no Lyme disease in Santa Cruz County. The doctor Elizabeth consulted refused to even look at the tick. No matter that we found several other women with Lyme disease picked up from tick bites in Santa Cruz County. So far as our doctors were concerned, she was not allowed to have it.

The significant thing about it was that Lyme disease, and/or the treatment she eventually received for the disease, had a crippling effect; it destroyed the cartilage in her right knee. But to get treatment, she had to be taken by our daughter Marlene to Ukiah, where we knew a Lyme disease specialist did exist. He identified the tick, and the disease symptoms, and started her on antibiotics, referring her to a San Francisco specialist for further treatment.

It was March 29 when she was bitten, and not until October 3 that we gave up on the local doctors and went to Ukiah. That was plenty of time for the spirochete to get dug in. For the next year she received regular treatment monthly from the S.F. doctor, along with checkups from a local Santa Cruz M.D. who didn't really believe in Lyme disease either.

At first Elizabeth was not too badly handicapped by her bad knee. She could still drive her car and get around with a cane. But her condition worsened over time and she had to make use of a walker. It was so typical of her approach to life that she put wheels on the rear legs of the walker, converting it to a "high-speed walker," the only one in town.

Hoping to get a better look at the biosphere reserve in Mendocino and the Sonoma and Marin coast—in my capacity as board member of the Golden Gate reserve—I toured the reserve with Elizabeth in 1994 after stopping off to visit Laurie Wayburn in Boonville. That year also the board of the World

Wildlife Fund held its annual meeting in San Francisco. Since I was a member of its national council too, I helped show them the biosphere reserve, at least the Marin County part of it. Russ and Aileen Train were there, which led to a pleasant reunion with them.

But the year was to end badly. From October 6 to 13 Elizabeth had a series of tests, followed by an examination on November 4. By November 14, it was decided by Dr. Holbert and Dr. Albritton, the surgeon, that she had colon cancer. They operated on November 29 and she was home recovering by December 5. On the 20th we were told that all was OK. Elizabeth had bounced back in her usual style. In the midst of it all, however, my old friend Dick Cooley of Alaska fame died suddenly on November 18.

The first half of 1995 seemed bright and reasonably happy for us. But then in late July, new tests indicated more trouble. A CT scan on September 12 showed the cancer had spread to her liver. We went to UC's Moffitt Hospital in San Francisco for an operation on October 6. Once again, Elizabeth fought back and seemed to recover quickly.

Then on a horrible night, January 30, 1996, Elizabeth fell and broke her hip. Once again to Dominican Hospital in Santa Cruz. Once again she fought back, and with the aid of physical therapists she was again able to move around on her walker. She was home by February 22. Then on March 14 we were called in and went to see Holbert. The cancer had spread widely. It was inoperable, but there was still some hope from chemotherapy. We went through that painful process, but I was told by Holbert that the situation was pretty hopeless. He gave her three months to a year to live, with the likelihood it would be closer to three months. On September 8 she fell down and could no longer walk. She was totally bedridden until October 15, when she died.

To say this happened is to avoid talking about the sheer hell we were all going through. Somehow Elizabeth and I had not envisioned the death of just one of us but had always assumed we would go out together, as we had done so many things together in our fifty-three years of married life. Instead there was this long, painful process. I could not have taken care of Elizabeth at home by myself, but when it had all become desperate in the spring of 1996, our oldest daughter, Sandra, moved in with us, to be followed by Marlene, who moved from Oakland to Santa Cruz. They took over their share of the home care. We were aided by cheery visiting nurses, initially from the Visiting Nurse Association and later, when things got hopeless, by nurses from hospice.

Elizabeth was heavily dosed with morphine, which relieved the pain but made it difficult for her to talk. I felt so damned helpless. I would have done anything for her, but nothing I could do really helped. Worst was the last night. I was exhausted but still wanted to stay awake and available. I think I finally fell asleep around midnight. When I awoke with a start at 3:00 A.M., she was gone. I mean gone. Her body was there, but she was not.

So I was totally depressed. Of course I contemplated suicide. But from everything I had read and believed, this would be entirely wrong and might actually prevent us from getting together for a long, long time. My daughters made a lot of difference. The losing battle we had waged to help Elizabeth recover and be well again hit us all equally hard. They needed me and I needed them. Our friends rallied around and helped us put on a memorial reception. My friends in the biosphere reserve and the university all helped. Peggy Lauer said to me, "You don't know how lucky you have been, to be with the woman you loved for fifty-three years." It was true. We had so many amazing adventures all over the world. Few have been so fortunate.

Both Elizabeth and I believed strongly in reincarnation and the laws of karma. We had talked about where we would like to be born in our next life. But of course now I was uncertain. I wanted to know she was there, somewhere. So in the interval between sleeping and waking, when you seem to have a foot in both worlds, I saw her again. She was once again young and beautiful and she told me not to worry, we would be together again. In fact, the last thing she had said to me when she was dying was, "Remember we are one."

Still I was a worrier. Maybe we did not know what the rules were. Maybe she could not wait for me on the other side and I would have to spend forever in a search for her through countless universes. Fortunately, a good friend of my Sandra's, Cynthia Lester, is a psychic who can go into a trance state and search for answers to any questions about past lives or future prospects. So I consulted Cynthia. She didn't have to search, she said, Elizabeth was right there.

Elizabeth told us about dying, how she had kept on fighting because she didn't want to leave us. She said she had been like a tree frog hanging on with all four sticky feet, so she was hanging onto life finally with her last foot and then suddenly her foot let go and she found herself shot like an arrow to a beautiful place, feeling surrounded by love. Her friends and relatives who had died were with her to welcome her. Now she said it was like being in a high place looking over the world. She could "paint" any landscape from the rain forest to the Himalayas, and she was busy. She told me we would be together in that place outside time where there was no past or future. Not to worry about future lives; when I got to where she was I would never want to leave. As for now, I should be getting on with my life. My life in this body was limited, whereas when I joined her time would be infinite. She was busy and I was not to worry. She could not "baby-sit" me but looked forward to our being together again.

There was more to it than that, a full two hours of hearing about how things had been in past lives. I asked about whether or not we had been together in the south of France. When Elizabeth and I had visited the Languedoc, we had felt it was a bad place for us. Cynthia had never heard of the Cathars, or the massacre of the Cathars or Albigensians in the thirteenth century, and was shocked and pained by what she saw during the session with me. Yes, Elizabeth and I had been there and had been killed, along with most of the other Cathars.

For me this was convincing. Cynthia never knew Elizabeth and had no way of knowing her ways of speaking. Yet when she was repeating directly what Elizabeth told her there was no question in my mind about who was speaking. It was my wife. We had not accepted the "til death do you part" bit of the marriage ceremony. We were married forever. This time around, Elizabeth pointed out, she was the follower. "But don't think it will always be this way!" Later, in another session with Cynthia, I complained that Elizabeth had always been ahead of me spiritually, and now she kept getting farther ahead all the time. "What do you expect?" Cynthia said. "She's a woman!"

I don't know whether I might be a woman in my next incarnation. Nobody has told me yet. But whatever happens I will have to accept it. I just hope it doesn't take another world war to bring Elizabeth and me together again. Maybe next time she will be "the girl next door."

# BIBLIOGRAPHY

Berg, P., and R. F. Dasmann. 1976. Reinhabiting California. *The Ecologist.*

Black Elk. 1972. *Black Elk Speaks.* As told through John G. Neihardt. New York: Pocket Books.

Borgstrom, Georg. 1965. *The Hungry Planet.* New York: Macmillan.

Brower, Kenneth. 1976. *Micronesia: Island Wilderness.* San Francisco: Friends of the Earth.

Brown, Dee. 1971. *Bury My Heart at Wounded Knee.* Reprint, New York: Holt, Rinehart and Winston, Bantam Books, 1976.

Burnette, Robert, and John Koster. 1974. *The Road to Wounded Knee.* New York: Bantam Books.

Dale, Tom, and Vernon Carter. 1955. *Topsoil and Civilization.* Norman: University of Oklahoma Press.

Darling, Frank Fraser. 1937. *A Herd of Red Deer.* London: Oxford University Press.

———. 1947. *Natural History in the Highlands and the Islands.* London: Collins.

———. 1955. *West Highland Survey: An Essay in Human Ecology.* London: Oxford University Press.

———. 1970. *Wilderness and Plenty.* Boston: Houghton Mifflin.

Darling, Frank Fraser, with J. Morton Boyd. 1969. *The Highlands and the Islands.* London: Collins, Fontana New Naturalists.

Dasmann, R. F. 1964. *African Game Ranching.* Oxford: Pergamon Press.

———. 1965. *The Destruction of California.* New York: Macmillan.

———. 1975. Difficult Marginal Environments and the Traditional Societies Which Exploit Them. *News from Survival International,* November 11.

———. 1976. "Future Primitive": Ecosystem People versus Biosphere People. *Co-Evolution Quarterly* (fall): 26–27, 29.

———. 1984. *Environmental Conservation.* 5th ed. New York: John Wiley and Sons.

———. 1988. Toward a Biosphere Consciousness. In *The Ends of the Earth,* ed. Donald Wurster. Cambridge: Cambridge University Press.

Dasmann, R. F., with John Milton and Peter Freeman. 1973. *Ecological Principles for Economic Development.* London: John Wiley and Sons.

Ehrlich, Paul R. 1968. *The Population Bomb.* New York: Ballantine Books.

Eichelberger, Lt. Gen. Robert L., and Milton Mackaye. 1949. "Take Buna . . . or Don't Come Back": Our Bloody Jungle Road to Tokyo. *Saturday Evening Post,* 17–19, 104–5, 108–11.

Farvar, Taghi, and John Milton, eds. 1972. *The Careless Technology.* Garden City, N.Y.: Natural History Press.

Graham, Edward. 1944. *Natural Principles of Land Use.* London: Oxford University Press.

Hawkes, Jacquetta. 1953. *Man on Earth.* New York: Random House.

Holdgate, Martin. 1999. *The Green Web: A Union for World Conservation.* London: IUCN; Earthscan.

Humboldt, Alexander von. 1849. *Aspects of Nature in Different Lands and Different Climates.* Philadelphia: Lee and Blanchard.

International Union for the Conservation of Nature and Natural Resources (IUCN). 1974. *United Nations List of National Parks.* Morges, Switzerland.

Jacks, G. V., and R. O. Whyte. 1939. *Vanishing Lands.* New York: Doubleday Doran.

Longhurst, W. A., S. Leopold, and R. F. Dasmann. 1952. A Survey of California Deer Herds, Their Ranges and Management Problems. *California Department of Fish and Game, Game Bulletin* 6, Sacramento.

Macy, Joanna. 1985. *Dharma and Development.* Rev. ed. West Hartford, Conn.: Kumarian Press.

Maillard, Joseph. 1930. *Handbook of the Birds of Golden Gate Park, San Francisco.* San Francisco: California Academy of Sciences.

Martin, Calvin. 1978. *Keepers of the Game.* Berkeley: University of California Press.

Martin, Paul. 1973. The Discovery of America. *Science* 179: 969–74.

McLuhan, T. C., ed. 1971. *Touch the Earth.* New York: Promontory Press.

Nash, Roderick. 1967. *Wilderness and the American Mind.* New Haven: Yale University Press.

Osborn, Fairfield. 1949. *Our Plundered Planet.* Boston: Little Brown.

Ricketts, E. F., and J. Calvin. 1939. *Between Pacific Tides.* Rev. Joel Hedgpeth. 3d ed. Stanford: Stanford University Press.

Schoenherr, Allan, with C. R. Feldmeth and M. J. Emerson. 1999. *Natural History of the Islands of California.* Berkeley: University of California Press.

Sears, Paul. 1935. *Deserts on the March.* Norman: University of Oklahoma Press.

Stewart, George R. 1948. *Fire.* New York: Random House.

Taber, R. D., and R. F. Dasmann. 1958. The Black-Tailed Deer of the Chaparral. *California Department of Fish and Game, Game Bulletin* 8, Sacramento.

Thomas, W. L., ed. 1956. *Man's Role in Changing the Face of the Earth.* Chicago: University of Chicago Press.

Turnbull, Colin. 1972. *The Mountain People.* New York: Simon and Schuster.

Udvardy, Miklos. 1975. A Classification of the Biogeographical Provinces of the World. International Union for the Conservation of Nature, Occasional Paper 18. Gland, Switzerland.

Vogt, William. 1948. *Road to Survival.* New York: William Sloane.

Wallace, Alfred Russell. 1876. *The Geographical Distribution of Animals.* London: Macmillan.

———. 1892. *Island Life.* London: Macmillan.

Weaver, J. E., and F. E. Clements. 1938. *Plant Ecology.* 2d ed. New York: McGraw Hill.

Yocum, Charles, and Raymond Dasmann. 1957. *The Pacific Coastal Wildlife Region.* San Martin, Calif.: Naturegraph.

# INDEX

Page numbers in italics refer to plates.

Compositor: Integrated Composition Systems
Text: 11.5/15 Granjon
Display: Granjon
Printer and binder: Thomson-Shore, Inc.